Airmanship

This book takes readers beyond the technical manuals and cockpit procedures to explore aviation as a lived experience. Based on years of ethnographic fieldwork, this book examines the culture, mindset, routines, teamwork, and professional identities that define airmanship. Offering a glimpse into the lives of pilots and flight attendants, they emerge not as anonymous operators of a system, but as individuals negotiating judgement, responsibility, and trust in an environment where precision and safety are paramount.

This title offers a compelling exploration of the people, practices, and philosophies that make aviation possible. Drawing on over 1,200 hours of jump seat observations, conversations with pilots and crew, and first-hand experiences in high-performance environments, the book offers an inside view of how aviation professionals make sense of their world. It shows how flight attendants manage safety and care in the cabin, how pilots read and interpret the sky, their aircraft, and themselves, and how crews cultivate a shared professionalism that sustains aviation. Through interesting stories and detailed analysis, readers will gain a deeper understanding of the human dimensions of aviation, with insights that enhance professional practice and personal appreciation for the craft.

Airmanship: Ethnography of Aviation Safety is ideal for human factors specialists, pilots, aviation students, safety professionals, and readers curious about the world behind the cockpit door. It is not only a study of aviation; it is a study of how people learn, act, and create safety together.

Vegard Nergård is a Professor of Aviation at UiT The Arctic University of Norway. With over 16 years of experience in aviation education, he was central in establishing UiT's master's programme in Aviation in 2015. He currently serves as Head of the Aviation and Drone Research group at the Department of Technology and Safety (ITS), UiT, where he oversees one of Norway's leading academic environments for aviation studies. His research focuses on airmanship, aviation safety, and pilot training, with a particular emphasis on Arctic conditions and environments.

Airmanship
Ethnography of Aviation Safety

Vegard Nergård

CRC Press
Taylor & Francis Group
Boca Raton London New York

CRC Press is an imprint of the
Taylor & Francis Group, an **informa** business

Designed cover image: Christian Aasen Stratton

First edition published 2026
by CRC Press
2385 NW Executive Center Drive, Suite 320, Boca Raton FL 33431

and by CRC Press
4 Park Square, Milton Park, Abingdon, Oxon, OX14 4RN

CRC Press is an imprint of Taylor & Francis Group, LLC

Library of Congress Cataloging-in-Publication Data
Names: Nergård, Vegard, author
Title: Airmanship : ethnography of aviation safety / Vegard Nergård.
Description: Boca Raton : CRC Press, 2026. | Includes bibliographical
references and index.
Identifiers: LCCN 2025044028 | ISBN 9781041211655 hardback | ISBN
9781041197898 paperback | ISBN 9781003725862 ebook
Classification: LCC TA163.N47 2026
LC record available at https://lccn.loc.gov/2025044028

ISBN: 978-1-041-21165-5 (hbk)
ISBN: 978-1-041-19789-8 (pbk)
ISBN: 978-1-003-72586-2 (ebk)

DOI: 10.1201/9781003725862

Typeset in Times
by SPi Technologies India Pvt Ltd (Straive)

Dedication

Dedicated to my father, my wife, and daughters.
Thank you, all pilots and crews, who took the time to teach me.

Contents

About the Author

Vegard Nergård is a Professor of Aviation at UiT The Arctic University of Norway. With over 16 years of experience in aviation education, he established UiT's master's programme in Aviation, which launched in August 2015. Dr. Nergård's research focuses on airmanship, aviation safety, and pilot training methodologies, particularly in Arctic environments. As Project Manager for UiT's aviation studies, he has helped develop the university's aviation programme, which graduates approximately 24 students annually with bachelor's degrees in aviation and CPL certificates, and his interdisciplinary approach bridges practical aviation knowledge with academic research.

Acknowledgements

This book was written at the Scott Polar Research Institute and at Christ's College, University of Cambridge. I am grateful for the time and the quiet I was given there.

Preface

Across cultures and throughout history, the sky has been a source of awe and mystery, a space where human imagination soars alongside the birds that inhabit it. Aviation, with its flying machines that defy gravity, is far more than a technical marvel. It represents a deeply respected human activity that resonates with high human aspirations. From the myth of Icarus to the pioneering flights of the Wright brothers and the cutting-edge technologies of today's aviation, the desire to conquer the skies reflects a universal pursuit of freedom, exploration, and transcendence.

This book approaches aviation from an ethnographic perspective, focusing on the culture of airmanship and the skills, values, and experiences that define the aviator's craft. The act of flying is far more than journeys through the air. It is an existential undertaking that challenges us to reflect on our relationship with nature, technology, and the limits of human achievement. From the artistic visions of da Vinci to Sir George Cayley's groundbreaking principles of aerodynamics, the transition from myth to science reveals a rich tapestry of cultural and technological evolution that continues to shape the way we engage with flight.

Through some of the practices and philosophies that shape the world of aviation, this book deals with airmanship both as a technical discipline and a cultural practice. The aviator's journey, from training to advanced skill in the cockpit, is a deep personal and professional transformation, shaped by a shared commitment to safety, teamwork, and communication. Through this lens, airmanship becomes not only the mastery of flight but a way of life, steeped in responsibility and the constant negotiation between risk and reward.

Aviation's profound influence extends far beyond individual pilots or even the industry itself. Aviation takes an increasingly dominant role in redefining the world, shrinking distances and connecting people, cultures, and economies in ways that were once unimaginable. Air travel has fostered a global sense of unity, enabling cultural exchange, trade, and international collaboration. Yet, its impact is not without complexities. Its history is deeply intertwined with military developments – reconnaissance balloons, the rapid advancements in aircraft during global conflicts, and, most recently, the reliance on unmanned aerial vehicles (drones) in modern warfare, such as in the ongoing war in Ukraine. These developments highlight the dual nature of aviation as both a tool for progress and a potential instrument of human disasters.

Beyond geopolitical concerns, aviation also poses significant environmental challenges. The ecological impact of air travel demands urgent attention and sustainable solutions, as the need for global connectivity must be balanced with environmental stewardship. The industry's carbon footprint underscores the tension between technological progress and the responsibility to protect and save our planet for future generations.

Aviation also embodies intrinsic risks. The men and women who operate aircraft confront these dangers directly, demonstrating remarkable courage, resilience, and the ability to embrace uncertainty. Whether in commercial aviation, air ambulance, special operations, or space exploration, these individuals willingly engage with the precarious nature of human life, accepting the unknown in pursuit of advancement and connection.

My book delves into the human dimensions of aviation, examining how interpersonal interactions, team dynamics, and communication play pivotal roles in the success and safety of flying. Airmanship, at its core, is about more than technical proficiency. It is about building trust, fostering collaboration, and navigating the unpredictable, all within a framework of shared responsibility and respect for the power of flight. Exploring these themes is to illuminate the meanings embedded in the practice of aviation and the cultural significance of those who dedicate their lives to mastering the skies.

My experience in aviation spans almost 30 years, during which I have been observing the industry from different angles. I have been involved in selecting, training, and working with aircrew across civilian and military operations. Drawing on these experiences of long-term fieldwork, I present some of the observed challenges and analyse them from social and psychological perspectives.

The study of practices in everyday life in the sky over time and varying conditions is termed "fieldwork." Its classic form is the long-term residence in a community or institution by the scientific observer to document and try to understand the essence of life and human activity in a particular field. In anthropology, fieldwork was established as long-term studies carried out mostly by European and American anthropologists and academics from the early 20th century. These studies in non-Western societies led to a heated debate on the perspectives through which a foreign culture can be studied and by what criteria those perspectives are valid and acceptable.

The early cultural studies were conducted at the writing desks in Western universities. The British anthropologist Edmund Leach nicknamed their practitioners "the stay-at-home theorists" (Leach, 1982). As the need for long-term presence in the ethnographic field developed, the opinion of the cultural stranger increasingly became ordinary humans in field studies. It replaced the Western ethnocentric opinion and was most clearly rejected by Clifford Geertz in his celebrated lecture and article: "From the Native's Point of View" (Geertz 1974). His position on behalf of his subject was not optimistic, but he advocated the role of the fieldworker as that of the interpreter, translator, and writer (Geertz, 1984).

The history of cultural studies is highly relevant for understanding other fields, as well as how pilots and members of the crew act, experience, and develop meaningful relationships. Flight operations must be understood close to the perspective of the operating actors. To get closer to the order of flight operations, long-term fieldwork has been necessary. Nevertheless, it is as challenging as any other form of fieldwork to get close to "the native`s point of view" in the Geertzian sense. My fieldwork has included a long period of flying frequently with experienced pilots, sitting in the jump seat[1] for over 1200 hours logged in various aircraft, and during different flight operations[2]. The purpose has been to learn, first-hand, how collaboration unfolds and how problems are solved under various operative conditions. This research has focused specifically on the experience of human collaborative action within high-performance environments, the means by which people solve problems collectively under operative conditions, and the explanations pilots and crew give of the challenges and choices made during a flight.

From qualitative interviews, fieldwork participation, and a decade of work as a Crew Resource Management (CRM) instructor, I have seen pilots and flight instructors who only partly view communication skills as central in airmanship. This stands in contrast to the findings of many accident reports, where breakdowns in communication are most often listed as the most decisive factors. The United Airlines Flight 173 accident in 1978 is often described as one of the events that triggered the very idea of CRM in commercial aviation. Against this backdrop, I very early on designed a qualitative research project where pilots were asked to share their thoughts and reflections on the subject.

Between 1996 and 2025, I have met, talked with, interviewed, and been in contact with around 500 pilots from different airlines. Their backgrounds vary. Some are air force pilots. Some received their initial training in the military before moving into civilian operations. Most, however, have completed a civilian training pathway and work in commercial airlines. Nearly all hold an Airline Transport Pilot Licence (ATPL), the highest level of qualification. Experience levels differ greatly. One pilot had logged 28,000 flight hours, while another had only 20.

The interviews took place in many different settings: in simulators, in cockpits at cruise altitude, in offices, and at airports. Most of the interview material presented in this book has been translated from a Scandinavian language into English. Language is never neutral, and in aviation, it is deeply tied to culture. Perceptions of safety and good airmanship are, to some extent, shaped by cultural differences.

UNVEILING LAYERS THROUGH ETHNOGRAPHY

To understand the multifaceted layers contributing to successful and safe flying, I have used various methodologies, although this text is almost entirely based on ethnography and observation conducted over many years. Fieldwork is a tool that enables a researcher to delve into the circumstances at a field site. The ethnographic approach offers a position close to flight operations, and safety matters can be examined comprehensively. At its core, ethnographic research methodology allows the researcher to come close to the "native" community of aviation. It opens for close attention to adjusting the research approach to the environment in question. When I started my research years ago, my intention was to observe the people, behaviours, practices, and dynamics within the business. The approach was intended to capture the lived experiences and perspectives of individuals and communities by observing their everyday lives. I hoped to understand flying from within and capture the layers of meaning, social interactions, and cultural dynamics that shape the lives of pilots and their crews. By participating in the everyday lives and practices of pilots, air traffic controllers, ground crew, and others, I hoped to come closer to their experiences, routines, challenges, and decision-making processes.

My methodological aim was to identify and explore the factors that contribute to flight safety, understanding how pilots and crew interact to ensure the secure movement of an aircraft from point A to point B. I had the opportunity to step into the cockpit and observe the professional activity at the nerve centre during flight operations. By shadowing pilots during pre-flight briefings, cockpit preparations, take-offs, landings, and other critical moments, I gained access to the behavioural dynamics occurring between pilots, the exchange of critical information, and the application of the operating procedures. It was possible to identify how communication, cooperation, and shared situational awareness interact to ensure a safe flight.

Physical presence on the flight deck allowed me to map the intricate web of communication there - between pilots and air traffic controllers, the ground crew, and how safety protocols are ingrained in the routines and behaviours of aviation professionals. Safety briefings, risk assessments, and incident reporting form the backbone of collectively creating a safety-conscious environment during a flight.

How pilots handle adverse weather conditions, technical glitches, or unforeseen events appeared from the jump seat as a relationship between pilots, cabin personnel, passengers, and the aircraft. How pilots observe and interact with flight instruments, automated systems, and navigation aids requires some familiarity with what they do. Another challenge is to understand the human–technology interface in flight safety. By observing flight simulations, training sessions, and debriefings, I came closer to how pilots and instructors acquire and refine their skills. The pilots and instructors I met over the years have served as a lens through which I have studied the skills needed to form the tapestry of flight safety.

My research draws inspiration from Clifford Geertz's conceptual framework, particularly his distinction between "experience-near concepts" and "experience-distant concepts," to draw closer to how pilots perceive their attainment of safety (Geertz, 1983). In this context, the key element is whether the pilot community possesses "experience-near" modes of seeing and expressing safety, and how they achieve it. The book aims primarily to adopt a pilot-centric perspective and leverage their "experience-near" conceptualizations of safety.

Historically, the aviation industry has witnessed a shift in perspective on safety. CRM has compelled pilots to embrace behaviours that conform to safety-oriented norms and standards. In aviation management, discourse on flight safety often attributes value to the systemic elements rather than the actions of individual pilots. The balance between experience-based judgement and the increasing reliance on advanced technology is a matter of debate and concern in the aviation community.

The Norwegian philosopher Jakob Meløe expounded a distinction between three positions of knowing, metaphorically as the gaze of someone trying to understand a reality with which we are not familiar: "The knowledgeable gaze," "the unskilled gaze," and "the dead gaze" (Meløe, 1979). The three concepts provide the ethnographic approach that influences the researcher's perception and interpretation of the research issue in question.

The knowledgeable gaze belongs entirely to the insider with experience and familiarity with a certain culture or practice. This position enables one to recognize subtle nuances, interpret actions, and understand connections without further investigation of the things already known to other insiders. The unskilled gaze belongs to the outsider, aware of his position and the existence of an inside world to which he or she has no access. The unskilled do not deny or ignore the reality with which he/she want to be familiar with. That is the situation of a researcher who studies a foreign community or practice. Obtaining entry depends on cracking the code and seeing the inside of the subject group. The dead gaze belongs to those merely seeing their own world and hence denying the existence of any other. To Professor Meløe, the gaze of the unskilled is the position to strive for in research. It is bound to the awareness of not knowing and the willingness to learn from the experienced and be trained.

Meløe's metaphors are most relevant for research into flying and safety. An additional challenge is the gaze needed for a research environment replete with advanced technology and skilled operations. This requires the ethnographer's willingness to identify the gaps in their own understanding of the research subject.

Ethnographic research is, for the most part, a long learning process. Being aware of this in advance may turn observations into knowledgeable insights and avoid the effect of a dead gaze. My aim has been to take a humble and open perspective on the ethnographic challenge in this study.

Vegard Nergård
Cambridge/Tromsø, 2023/2025

NOTES

1 A *jump seat* in an aircraft cockpit is an extra seat, typically used by authorized personnel such as off-duty pilots, trainees, or inspectors. It is usually foldable and located behind or beside the pilots' seats. Although it does not offer the controls as the main seats, it allows the occupant to observe and communicate with the pilots during flight.

2 I have worked as a CRM and Human Factors instructor in two different airlines. I have read much of the PPL and CPL theory, but probably not understood most of it.

1 Preparing the Flight

Captain Ove and his flight crew invited me to join them in the breakfast room at the hotel. He had landed in Oslo late the night before. Yet, he looked surprisingly alert at 05:30. After a quick breakfast, we headed to the airport.

As we left the breakfast table, Ove spoke about how the industry had changed during his career. His words carried echoes of an earlier time, when flying was still marked by glamour and exclusivity.

Old photographs from the 1930s show cabins that looked more like lounges than machines. Men in suits and women in evening gowns sat in wide leather seats, drinking champagne from real glasses while a chef carved meat at the aisle. Flying was not simply a way of moving from one place to another. It was a theatre, reserved for the few who could afford the ticket. Figure 1.1 captures this earlier cabin culture - service as performance, space as theatre, and flying as an event rather than a routine. Ove never lived that world himself. What he knows comes from older colleagues, often in crew rooms late at night or over coffee on long layovers, when stories about

FIGURE 1.1 SAS DC-8-33. First Class. 1969. Interior of cabin, service on board by chef and air hostess.

(Copyright SAS/SAS Museet.)

DOI: 10.1201/9781003725862-1

the "golden age" of flying were told with a mix of pride and irony. They spoke of passengers dressing for the flight as if it were a wedding. Ove smiles when he recalls their words. Even in his own early years, he says, there were still traces of that formality in the way people behaved on board.

Through the decades, the theatre began to close in. Jet engines, larger fleets, and cheaper tickets filled the seats with new passengers. The cabin shrank, the seats moved closer, and service was reduced to something efficient rather than grand. Ove watched it change around him. "People stopped dressing up. Flying was no longer a celebration. It became transport. But the responsibility in the cockpit was the same."

The rise of low-cost airlines sharpened the change. Today, Ove works in that environment. The pace of the short haul is fast, turnarounds are tight, and the flight is judged by its punctuality more than anything else.

As we approach the airport, a heavy aircraft is taking off. The airline operating it is known for its revitalization of luxurious flying, returning at the other end of the market than what Ove's airline is known for. The airline sells privacy in first-class suites with doors and beds, champagne before take-off, and service designed for exclusivity. Ove looks up at the departing aircraft. He shrugs. For him, the essence is unchanged. "It is still about moving people safely. My job is to bring people home. The frame around aviation has changed, but the craft has not."

Thus, while air travel today is more inclusive and affordable than ever, the divide between luxury and low-cost aviation reflects a fundamental shift in the industry. Ove finished his thought with a smile. For him, the industry may change its appearance, but the essence of flying does not. At the airport door, he turned back to me and smiled, as if to say, enough talk about history, now we get on with the work.

1.1 AT THE AIRPORT

We left the hotel behind and stepped into the familiar rhythm of the airport morning. The light was brighter, the air more hurried, and the conversation between us slowed as the ritual of control approached. We arrived at the security check-in. At the security line, everything moved into the well-known pattern.

The officer recognised Ove at once. They had met many times, most recently yesterday. But recognition changes nothing. Shoes off, belt off, metal objects in the grey box. The ritual is repeated every morning, sometimes more than once a day.

"I suppose I have to take off my shoes again," Ove said.

"Yes, unless you bought new ones since yesterday," the guard replied with a dry smile, watching him closely.

Ove did not answer. He placed his shoes on top of the belt and pushed the box through the scanner. He rolled his eyes at me before stepping into the body scanner. No sound.

"Shock," he said with irony. The guard on the far side smiled.

"You know, that is how it is," she replied.

"Yes, I know. I am thrilled you are here. It makes our life safer." His voice was calm, almost knowing. A connection was made, irritation already gone.

Then it was my turn. The alarm went off. I patted myself down in confusion before realising my wireless earphones were still in my pocket. I handed the small white box to the guard with an apologetic smile.

"Put them in the tray and try again."

Ove checked his watch, then looked at me with a grin. "Amateur. Let's hope you have better luck for the rest of the day."

The checkpoint was not only about metal detectors and trays. It was part of the working day, a reminder that even those in command of the aircraft are subject to the same rituals as everyone else. Shoes back on, earphones stowed away, we walked up the stairs to the crew room where the real preparations began.

The first officer was already there. He and Ove shook hands like old colleagues. Their words in Norwegian were light but warm.

"It will be good to fly with you again. It has been many months."

"Totally agree. You always make me look good," the first officer smirked, glancing in my direction.

Ove introduced me. "You know this guy, he will be flying with us today. Head office has cleared it. You are fine with that?"

"Of course, I know him from before," the first officer nodded.

The meeting was short. Papers were checked, coffee was poured. Most of the preparation had already been done by the first officer. "The weather up north can be tricky, but not today," he said. "I even checked before I went to bed last night. Forecast looks fine, though you never know in Northern Norway."

On the table lay the briefing package, a stack of papers every pilot knows by heart – weather, route, and notices. To outsiders, it might seem like clutter, but it is, in fact, the script of the day. The first officer had already worked through much of it. For him, preparation was not about ticking boxes but about shaping a mental picture of the flight ahead.

The chief flight attendant entered. She was in her thirties, bright and professional, and greeted Ove warmly. "I saw you had requested this flight, and head office confirmed it. Good to see you again."

"Good to see you too," Ove replied. Their words were brief, but the tone was clear – respect and familiarity, contained within the frame of professionalism.

"See you in a few minutes," she said before disappearing back into the corridor.

These first minutes of the day were routine. Papers, coffee, handshakes, and short words carried a weight of their own. The atmosphere was light but disciplined, a reminder that trust in aviation is built long before the aircraft leaves the ground.

1.2 A CUP OF COFFEE WITH CAPTAIN OVE

We sat with coffee before the day began. The crew room was quiet at that hour, filled with the low hum of vending machines and the smell of fresh brew. I had known Ove for years, but he rarely spoke much about flying. Normally, our conversations circled around his children and the everyday things he cared about most. That morning, though, he let it drift towards his work.

"I don't think of flying as something grand," he said, stirring his cup slowly. "For me it is a craft. Something you refine over years. Hopefully my colleagues see me as someone they can trust. That matters more than anything."

He leaned forward as he spoke about his first officers. "If they see me make a mistake, or if something doesn't feel right, I want them to say it. No hesitation. That is how we stay sharp." He gave the same weight to the cabin crew. "They set the tone for the whole flight. If they are uneasy, we feel it up front. They are our eyes and ears back there. I want them to know they can always come to me."

There was no lecture in his words. He spoke plainly, as if these were obvious truths, but the steadiness in his voice gave them gravity. He never wasted words, and when he did speak, colleagues listened. Other pilots described him as the kind of captain who never made a fuss, no matter how stressful the situation became.

Ove carried himself with a quiet confidence. He was tall and broad-shouldered, but it was not his size that made him stand out. It was the way he filled a room without raising his voice. His presence was felt more in pauses than in speeches, in the way he looked at someone before answering.

When the talk turned to family, his expression softened. He smiled, leaning back in his chair, cup balanced loosely in his hand.

> I never wanted to be the pilot dad who is always away. After the divorce I cut my hours. I wanted to be there for birthdays, homework, all of it. Their mother and I make it work. The kids come first.

He laughed when I asked about hobbies. "My kids are my hobby. No sports, no side projects. Just them." His laughter was warm, but the words carried no irony. For him, it was the simple truth.

His way into aviation had not been direct. He once trained as an air traffic controller before deciding he wanted to be in the air, not on the ground. "I realized I would rather be the one flying than the one keeping others apart," he said with a grin. With support from his parents and a state programme, he became a pilot, and after more than twenty years, he still loved the moment of take-off. "It is freedom and responsibility at the same time."

Flying came at a cost, he admitted, but he carried the same mindset into his family life as in the cockpit. Stay ahead of the situation, spot the risks early, and adapt when things change. He did not aim for perfection, only to be present where it mattered most.

When asked to describe himself, he paused, eyes fixed on the table before he answered. "Reliable," he said simply. He did not explain further and did not need to.

As we finished our coffee, he leaned back once more.

> Aviation is more than a job. It shapes how you think and how you handle yourself. How you handle pressure. But at the end of the day, it comes down to the people you work with, and the ones waiting for you at home.

Listening to Ove, it was clear that flying was never only about machines or routines. His way of speaking revealed a culture where trust and responsibility are carried in small, everyday acts. In that sense, even a coffee break was already part of the flight.

1.3 PLANNING IN THE CREW ROOM

In commercial aviation, the crew often meets just an hour before the flight. Time together was short, but the preparations were shaped by rules and routines set by the authorities.

"Since you are here, we might as well show off our outstanding professionalism," the first officer laughed, his hand resting on the stack of papers. Then his voice turned serious.

> We do these briefings as thoroughly as possible. I have been in airlines where leaders did not really care about safety. Here it was different. Policies were not just politics. The leaders wanted us to do things as safe as possible.

On the table, the papers were spread out, thin sheets covered in numbers and marks: weather charts, the day's flight plan, and notices from air traffic control (ATC). To an outsider, they might have looked like clutter, but to the crew, they were the script of the flight. In some airlines, the package no longer arrived on paper but on a slim electronic tablet.[1] Yet the ritual was unchanged. Pilots still leaned over the same data, pointing with a finger or the tip of a pen, making small notes on the screen. Whether paper or digital, the rhythm of preparation remained the same (Figure 1.2).

Ove gave a short nod and summed up the briefing they had just had. "Weather calm at departure and arrival. Thirty-six thousand feet cruising, descent into Tromsø. Aircraft scheduled for two more sectors after us. Should be straightforward." His tone was even, matter-of-fact, but each sentence carried the weight of a decision already made.

Then came the quiet agreement about roles – who would fly and who would monitor. No big declaration, just a glance and a nod. That day, Ove would fly, and the first officer would monitor. On the next leg, they would switch.[2] It was part of the rhythm of the job, a way to share both the effort and the responsibility. The one flying stayed absorbed in the moment, hands on the controls during take-off and landing, eyes scanning instruments even when the autopilot was on. The one monitoring kept the broader view, talking to ATC,[3] checking the fuel, reading out the checklists, and ready to catch whatever the other might miss. The alternation between the roles was as ordinary as breathing. It kept the workload balanced, and it kept trust alive.

Much of this happened in near silence. They conferred in low voices, broke off mid-sentence, and continued with gestures. The trust between them was carried more by rhythm than by words.

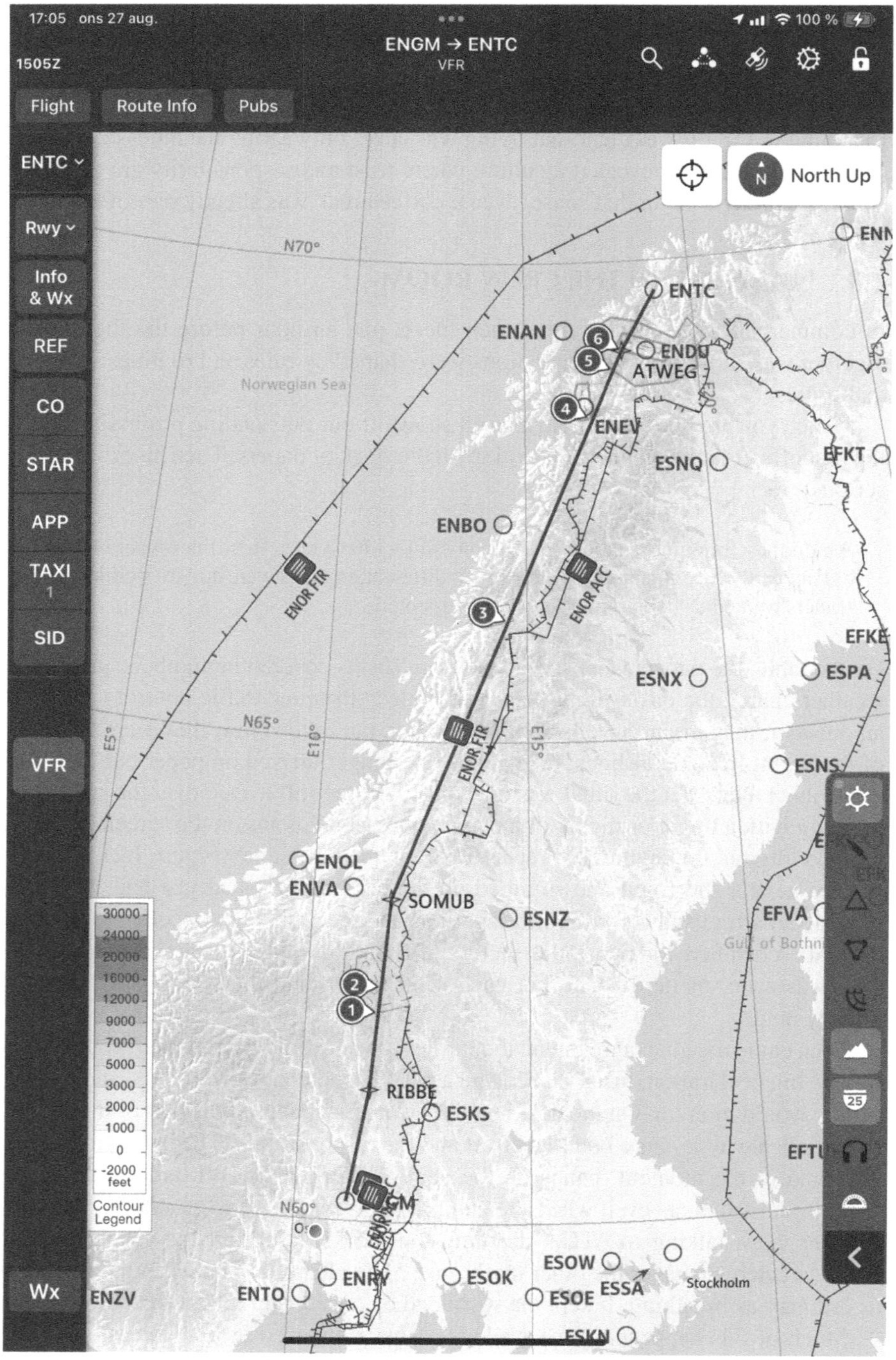

FIGURE 1.2 The planned route Oslo–Tromsø. The map of the Standard Departure from Oslo Gardermoen (an RNAV SID), displayed on the crew's Electronic Flight Bag (EFB).

(Snapshot: Vegard Nergård.)

Ove returned with coffee, setting a cup down by his colleague without comment. It was a small act, but it carried respect. Not the kind written in manuals, but the kind lived in routines.

The crew room itself hummed with ordinary sounds. Printers spat out flight plans, doors opened and closed, and the coffee machine hissed. Around the table, though, the atmosphere was focused, almost heavy. For the passengers boarding later, it would look like a simple flight north. But here, in the quiet rituals of briefing, the flight was already taking shape.

Ove turned to me, half smiling. He must have sensed I wanted the unsaid to be said aloud.

> After departure from Oslo, from runway 19L, we follow a standard departure route that takes us up over Eastern Norway. Then we level off. There's usually time for a coffee and maybe a quiet joke. But then, once we start the descent into Tromsø, everything picks up again. New terrain, new procedures, and often a different kind of weather. Low light. Fast changes. You stay ahead of it. That's the job.[4]

What looked like paperwork and routine from the outside was, in fact, the foundation of trust. In those short minutes, the crew rehearsed a pattern of cooperation that would follow them into the cockpit. It was here, in the quiet preparation, that safety began.

1.4 THE WALK TO THE AIRCRAFT

We left the crew room together, the briefing finished, but the day was only beginning. The corridor smelled faintly of coffee and jet fuel drifting in from the gates. Bags rolled behind us, shoes tapped on the polished floor. No one spoke much. The light conversation of the crew room gave way to a quieter focus as we walked through the terminal.

At the gate, the uniform set them apart. Passengers glanced at Ove and the first officer, some with curiosity, some with trust already written in their eyes. The flight attendants joined us, greetings short and professional.

We stepped through the last door, out onto the apron.[5] The sound was immediate. The hum of auxiliary power units, the distant roar of departing jets, and service trucks reversing with sharp beeps. The morning air was sharp, tinged with the smell of kerosene. The aircraft stood waiting, white fuselage streaked with traces of the night's weather, cabin lights glowing faintly through the windows.

Ove walked ahead, his stride steady, the first officer just behind him. There was nothing dramatic in their movement, but something in the way they carried themselves made it clear that the flight had already begun.

NOTES

1 An Electronic Flight Bag (EFB) is a digital service device, usually a tablet, used to replace paper-based documents such as flight plans, weather information, and manuals. Many airlines have adopted EFBs, but the practices of briefing remains largely the same.

2 Pilot Flying (PF) is the pilot who actively controls the aircraft. Pilot Monitoring (PM) supports by handling communication, checklists, and system oversight. On each leg of a multi-sector day, the roles are alternated. It is a practice designed to balance workload and sharpen attention.

3 A ground-based service responsible for managing and coordinating the safe and orderly flow of air traffic in the skies and on the ground.

4 On most departures from Oslo Airport (ENGM) to Tromsø (ENTC), aircraft use standard instrument departure (SID) routes from runway. The SID is designed to provide safe separation from other traffic and terrain during climb-out. Once established in cruise, the flight to Tromsø typically follows airways over Eastern and Central Norway before joining a STAR (Standard Terminal Arrival Route) into Tromsø, often leading to an RNAV or ILS approach to Runway 18 or 36, depending on the weather conditions. While these procedures are charted and rehearsed, local weather and lighting conditions in the north can shift rapidly. It is here that the value of preparation shows - especially when flying into complex terrain under low-visibility or winter conditions.

5 In aviation, the term apron refers to the area on an airport where aircraft are parked, loaded, refuelled, or boarded. It is part of the airside operations area, and crew access is strictly controlled.

2 Cabin Attendants

The cabin has always mirrored its time. In the early years, there were wide seats and meals served on china. Passengers dressed as if they were entering a private club. Later, there were smoking lounges, cocktail bars, and even beds. For a while, flying promised more comfort than most homes on the ground. Slowly, the rows grew tighter. The meals became simpler. The theatre was replaced by routine. Today, the cabin can be many things: cramped economy, quiet business comfort, or the rare private suite with doors and a bed.

This world was never shaped by passengers alone. For flight attendants, the cabin was a workplace where every redesign brought new tasks and new demands. The changing shape of the cabin through history has always carried new expectations for those who worked inside (Poirot 2024). What once had been a role of charm and glamour has become a role of vigilance, authority, and care.

Before the first passenger arrives, the attendants move through the cabin in practised silence. They touch the doors, pull at seatbelts, check oxygen bottles, and look into compartments. Each gesture seems ordinary, but each is tied to the most unforgiving of rules. In an emergency, the aircraft has to be emptied in 90 seconds.[1] Every announcement and every instruction about where a bag is stowed or whether an aisle is clear is shaped by this countdown. To passengers, the safety demonstration might look like a routine performance. To the crew, it is a preparation and a rehearsal for the one event they hope will never come.

Flight attendants are the first crew members passengers meet when they step onto the aircraft. The greeting at the door often sets the tone for the journey ahead. A warm smile can calm a nervous flier. A professional tone can ease tension in a queue of tired travellers: "It is everyone's job to ensure the passengers have an experience when flying with us," one chief cabin attendant told me (Interview with Lisa 2023). "While most passengers just want to reach a destination, for others it is about the journey itself. Our job is to make sure they arrive safely, and that they want to fly with us again."

Passengers often underestimate the complexity of the role. Some refuse to follow instructions. Some argue over luggage or electronic devices. Some dismiss the crew as waiters in the sky. What they rarely see is the quiet precision behind each task. Before boarding begins, the crew has already rehearsed scenarios of decompression, fire, evacuation, or medical emergencies. Each member knows their position, which door they are responsible for, what commands to give, and how to act if the cabin must be emptied in the dark or in smoke.

For the flight attendants, the flight begins long before the passengers arrive. They move quickly but deliberately through the cabin. The gestures are small but carry weight. Their first responsibility is safety. Each instruction to passengers - to stow a bag, switch off a device, or keep an aisle clear - carries a purpose.

DOI: 10.1201/9781003725862-2

2.1 A CABIN ATTENDANT

In the crew room before boarding, Lisa sat down with me, a mug of black coffee in her hand. She cradled it as if to draw warmth from it. Her eyes sparkled with a mix of pride and exhaustion. "It is a lifestyle," she said. "You either love it or you do not. I usually tell people I have the world's best worst job" (Interview 2023).

Her laugh was short, but it carried a note of truth.

> Every day starts the same way. Check-in. You meet your crew, the pilots, the cabin crew, the ground staff. Then comes the preflight briefing. It is more important than you would think. Because it is a formality. But it is here where we form our team. We go through safety procedures. We talk about possible threats such as bad weather, unruly passengers or medical issues. In our company we have an openness to ask and make sure everyone is fit to fly. If you are exhausted or unwell, you say it. Everyone depends on each other.
>
> *(Interview Lisa 2023)*

Once on board, the rhythm began at once. Safety equipment was inspected. Catering was checked. The cabin was prepared down to the smallest detail.

> When boarding starts we take our positions. Some at the doors, some in the aisles. We watch everything. The passenger greeting is important. It is partially about customer service and courtesy, but we actually do a quick evaluation and assess how people carry themselves, who might cause trouble, who might need help. The baggage, the seats, the mood in the cabin. Over the years that part has become more important, especially with the rise in unruly passengers.
>
> *(Interview 2023)*

Her tone changed as she spoke about passengers' expectations.

> People do not realise our main job is to make sure everyone can get off the plane if something happens. But what do they see. They see us handing out drinks. Every little instruction, like putting away your laptop or keeping the aisle clear, has a purpose. These small rules save lives. But passengers roll their eyes or argue. "I just need to finish this email," they say. They only notice us when something goes wrong.

Lisa explains with a huff.

She laughed softly and shook her head. "The hours are brutal. We flip between day and night, never really adjusting. Eating healthy? Forget it. Fitness? Impossible. Sometimes you grab a snack while waiting in line for the lavatory." She lifted her cup and paused:

> But the crew makes it worth it. Within minutes we are a team. We joke, we laugh. Passengers often notice how much fun we seem to have. That is the magic. When everyone is motivated and follows procedures, it is smooth sailing. When someone is not pulling their weight, the whole team feels it. Then the flight ends. Just as quickly as we came together, we scatter. New crew. New flight. And you do it all over again.

She described the safety routines with the precision of someone who had practised them a thousand times. Her hands moved slightly as she spoke, as though she was still performing the checks.

> When passengers board we are not just polite. We are scanning. Who makes eye contact, who avoids it, who looks nervous or intoxicated. Sometimes we escalate things deliberately on the ground because it is easier to handle with the doors open.

Her stories moved easily between the small and the dramatic.

> I have saved lives. I have done CPR at thirty-five thousand feet. I have held hands in turbulence, helped a child flying alone, comforted someone heading to a funeral. Sometimes it is smaller moments. A nervous flier. A smile that turns someone's day around.

She looked down at her coffee before adding, "You lose weekends, holidays, birthdays. Planning is almost impossible. Friends stop asking. If you are not comfortable in your own company this job is not for you."

Lisa paused, then gave a small smile.

> People think we are here to serve drinks. What they do not see are the emergencies we train for, the conflicts we defuse, the safety decisions we make in seconds. And at the end of the day, despite everything, I would not trade it for the world. I love the aviation industry.

2.2 FROM AIR HOSTESS TO FLIGHT ATTENDANT

The term *air hostess* was chosen to evoke a very particular image. Airlines wanted glamour in the cabin. They recruited young women who were unmarried and attractive, expected to serve with charm and elegance. The advertisements from the 1950s and 1960s showed smiling faces in pressed uniforms, promises of a career in the sky. The job was presented as a mixture of service and adventure, but always on the airline's terms. Weight and height were measured. Rules governed appearance. To fly as a hostess was to step into a role where the body itself became part of the company's image (see Figure 2.1).

They worked in heels, their uniforms cut to catch the eye, their weight checked by supervisors. They were told to smile even when turbulence made the service trolley shake. The cabin was a stage and their bodies part of the performance.

On board, the theatre continued. In the jet age, uniforms became brighter, sometimes playful, sometimes severe. Flight attendants posed in magazines, selling both destinations and the idea of modernity. One campaign asked passengers to "think of her as your mother." The irony of the message was clear. She was at once a figure of glamour and a figure of care, asked to embody both elegance and reassurance (see Figure 2.2).

The photographs now look dated, but they are a testament to how much the cabin reflected the ideals of its time. The hostess was there to smile, to calm, to be admired, and to make flying feel safe. The role was theatrical, and the demands were patriarchal. The young women carried their airline on their bodies.

FIGURE 2.1 Qualifications of the past.

(Unknown source)

Over the years, the image shifted. The word *stewardess* gave way to *flight or cabin attendant*.

The charm has never disappeared. On the contrary, it was no longer the only requirement. Safety has moved to the centre of the role. Training grew heavier. The crew became a visible authority in the cabin.

Today, the role is more complex still. Airlines recruit from many backgrounds, men and women alike. The focus is on competence. A new attendant is trained in

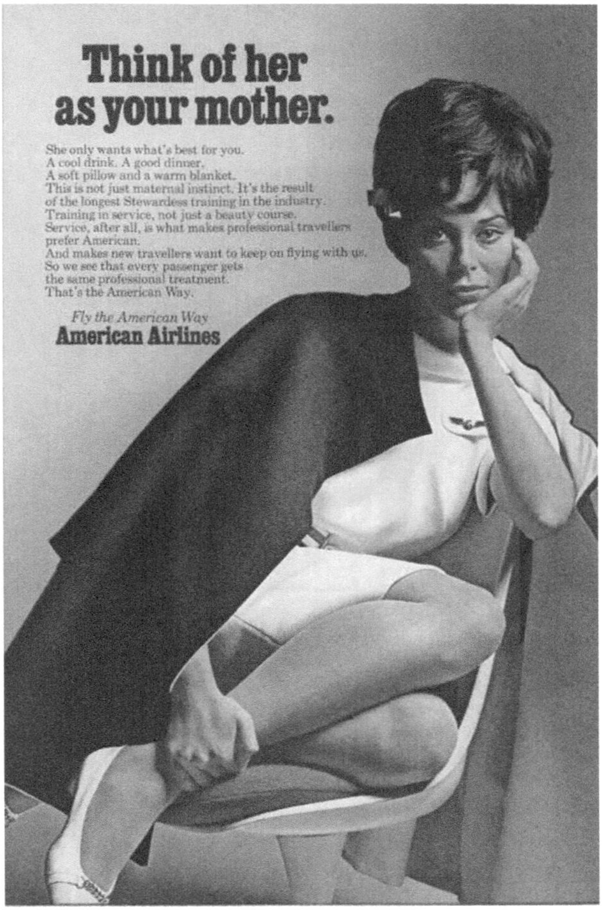

FIGURE 2.2 Think of her as your mother. Creating reassurance for safety and service.

Note: I found this picture in a paper: Lyths, P. (2012). "Think of her as Your Mother": Airline Advertising and the Stewardess in America, 1930–1980. *The Journal of Transport History*, 30(1), 1–21. https://doi.org/10.7227/TJTH.30.1.3 (Original work published 2009).

(Unknown source)

evacuation, firefighting, conflict management,[2] and medical care before ever serving a drink. They are still asked to smile, to be welcoming, but the job is no longer defined by that smile.

The uniform remains, but what it signals has changed. Where once it suggested glamour, it now signals authority, readiness, and trust. Passengers may still notice the smile first, but in an emergency, it is not the smile that matters. It is the authority in the voice, the strength in the hands that open the doors, the training that makes them act without hesitation (see Figure 2.3).

FIGURE 2.3 Modern cabin attending.

(Photo: Christian Aasen Stratton)

2.3 RECRUITMENT AND TRAINING TODAY

The pathway into the cabin today is very different from the one advertised in old magazines. Airlines no longer ask for height and weight or marital status. They look for competence. They search for people who can carry responsibility, who can stay calm when a cabin fills with smoke or when a conflict breaks out in the aisle.

The process begins long before the first flight. Applicants are interviewed, tested in groups, and observed as they solve problems together. The questions are not only about languages or previous work, but about how they read a situation, how they manage pressure, and how they speak when others are afraid. In training centres, new recruits sit in mock-up cabins with life vests and oxygen masks, doors that can be opened in seconds, and slides that must carry a full load of passengers to the ground.

The training is physical. They learn how to fight fire in a smoke-filled room, how to drag a body through a narrow aisle, how to secure a cabin in seconds. They repeat commands until the words carry authority. They practise the rhythm of pointing to exits, of checking belts, of moving through the cabin with eyes scanning left and right.

The training is also cultural. Airlines want crews who can greet a passenger in several languages, who can read politeness and offence across cultures, and who can carry both service and authority in the same voice. A good attendant must know when to smile and when to be firm, when to calm with words, and when to give a command.

The work has grown more diverse. Men and women enter from many backgrounds. Some arrive with degrees, some from hospitality, some from entirely different careers. What unites them is not how they look but how they act. The uniform no longer hides a body chosen for display. It marks someone who has passed tests that most passengers never see.

2.4 UNIFORM AND THE BODY

The uniform has always been part of the role. In the early years, it was designed to catch the eye. Today, it signals authority. The cut is more practical, but it still shapes how the attendants move and how passengers see them. Shoes are chosen for safety, yet they must still look polished. Jackets are adjusted for fit, but they must be worn even in summer heat.

The body is present in every shift. Flight attendants stand for hours, walk the aisles, bend to stow bags, and lean across seats to reach passengers. They serve meals in narrow spaces, balance trays in turbulence, and move quickly in emergencies. They handle heavy doors, check equipment, and sometimes restrain passengers. The work is physical, even when it looks effortless.

The uniform holds the body together in this work. It creates a common face for the crew. A passenger may not remember a name, but they remember the colour of the jacket, the cut of the hat, the impression of order. For the attendants, the uniform is both armour and a demand. It protects, it presents, and it reminds them that every gesture is visible.

Many attendants speak of the relief of taking the uniform off at the end of a long day. Without it, they can be anonymous again. They can walk through an airport unnoticed, their body no longer carrying the weight of the airline's image. Yet, the moment they put it on again, they step back into a role that carries trust and responsibility.

2.5 CLOSING REMARKS

The work of the cabin crew is often invisible. Passengers see the smiles and the service, but not the weight of the training or the decisions made in silence. Behind every calm announcement, there is practice repeated until it becomes instinct. Behind every gesture in the aisle, there is the memory of drills carried out in smoke and noise.

The role has changed over the decades. From glamour and display to safety and authority. From hostesses chosen for their appearance to attendants recruited for their judgment and resilience. Yet some things remain the same. The uniform still marks them out. The greeting at the door still sets the tone. The cabin is still their stage, though the performance has shifted.

To fly as a passenger is to step into a space that feels routine. To work as a flight attendant is to step into a space that must always be ready for the unexpected. It is here, between service and safety, between visibility and invisibility, that the role continues (see Figure 2.4).

FIGURE 2.4 Taking care of and attending.

(Photo: Christian Aasen Stratton)

NOTES

1 International safety regulations (EASA CS-25.803; FAA FAR 25.803) require that all transport aircraft must be capable of full evacuation within 90 seconds under test conditions.
2 See: Moira, P., Mylonopoulos, D., & Terzoglou, E. (2020). A Typology of Delinquent Behavior on the Aircraft. *Journal of Tourism and Leisure Studies*, 5(1), 29.

3 Acting at the Flight Deck (A Regular Flight)

Back in the crew room, the table was filled again. The captain and the first officer had returned, the cabin crew gathered close, and the tone shifted from casual to purposeful. The captain spoke of the weather along the route, the first officer added details of the flight plan, and the attendants shared what they knew from boarding: a family with small children and a passenger who needed special assistance. These fragments might sound small, but once the doors were closed, there would be no second chances to prepare.

When I asked later about cooperation, the crew described it as something constant, carried on in short exchanges throughout the day. A short message from the cockpit to the cabin to warn of turbulence so attendants could stow items and secure passengers. A quick call from the cabin to report a passenger who had fallen ill or become disruptive. "We communicate all the time," the captain explained. "That way decisions are made with the full picture." The first officer smiled and added, "In our company we say one aircraft, one crew. That is how we think about it." The captain nodded in agreement. "Yes, it is about teamness. We belong to the same operation" (Group interview 2023).

Soon afterwards, we stepped outside. The pilots moved ahead with steady steps, bags rolling behind them. The captain gave him a nod and led me up the narrow stairs into the cockpit of the Boing 737-800.

3.1 PREPARING THE FLIGHT DECK AND CABIN

Arriving at the cockpit,[1] both pilots take their seats. In front of them is the instrument panel, often referred to in aviation as the front panel, which houses essential flight instruments such as the blind flying panel and engine instruments. Between the captain's seat and the first officer's seat is another instrument console known as the pedestal.[2] This area typically contains key controls like the throttles, flaps, and communications systems. Above the pilots' heads, further instruments are mounted in the overhead panel (see Figure 3.1).

Any flight is divided into different phases. The session inside the crew room was the first part of the pre-flight and briefing element. Inside the cockpit, the sequence continues. The captain and first officer have defined patterns of work according to which specific operations must take place. The pilots each have their own sectors of the instrument panels to scan and perform tasks within. When starting the aircraft, the sequence of procedures follows a specific panel scan pattern. This pattern is described in the standard procedure manuals.

The captain proceeds with the "electric power" start-up procedures. At the same time, the first officer puts on his jacket and a reflective vest, finds his torch, and heads

DOI: 10.1201/9781003725862-3

FIGURE 3.1 The image shows the cockpit of a Boeing 737, one of the most widely used aircraft in commercial aviation. The cockpit is organized into three main areas. At the top is the overhead panel, where pilots manage systems such as fuel, electrics, anti-ice, and air conditioning. In front of the pilots is the main instrument panel, which displays essential flight data including altitude, speed, attitude, and heading, as well as autopilot and flight management system inputs. Between the two seats is the central pedestal, which houses the thrust levers, communication radios, and navigation controls. The captain sits on the left, the first officer on the right.

Photo: Vegard Nergård

out of the cockpit to conduct a walkaround, which is an external inspection of the aircraft. This is a mandatory pre-flight examination. The pilot inspects the aircraft for abnormalities in probes, vents, control surfaces, the landing gear, and brakes.

Meanwhile, the captain has started the electrical systems: "I do this all by heart, but everything is written down in our company procedures," the captain explained. The first officer returns after 10 minutes. Once he is back in his seat, the captain informs him of what he has done during his absence from the flight deck. The first officer double-checks the captain's actions. In that way, they cross-check that the system has been fed with correct information. The captain turns to me with a smile and claims:

> It is imperative that we double and cross-check. We are all humans. We do make mistakes. That's what humans do. The First officer checks that I have entered the right numbers into the computer. The computer is, on the other hand, completely stupid. If we put the wrong numbers into it, it returns shit. A saying among pilots is that dealing with computers is a shit in - shit out situation.

The pilots then proceed to the "stop/go briefing," which they do before every flight. The stop/go briefing is to prepare for a high-speed aborted take-off. If a malfunction or problem occurs during the take-off roll, the captain will call out "stop" to confirm a rejected take-off decision to indicate that he has control over the aircraft. A command of "go" confirms that he/she does not intend to reject the take-off. The captain explains: "When we pass 80 knots, we soon reach our decision speed and a 'go' state of mind. A 'stop' call will come if we see any alarms going off indicating engine or fire warnings, experience any severe damage, suddenly lose our engine thrust, receive a take-off configuration warning or any indication of a malfunction where there is doubt that the aircraft will fly safely."

The captain turns around to me:

> now we are in the process of doing a taxi and take-off briefing. Normally, this would be relatively short since the first officer is experienced. As you are with us today, I will give an extensive one and try to explain to you why I am saying what I am in the briefing.

After receiving a verification response from the first officer, the captain continued:

> In a normal operation, a good take-off briefing covers a variety of areas such as runway incursion avoidance, emergency procedures, weather, and pilot responsibilities. By formulating a plan and discussing certain contingencies, we prepare ourselves for the worst. In that way, we can recall the information more quickly if something should happen. In the case of anything happening on take-off, we do not have time to figure out what we are going to do. That is why we must prepare ourselves when we are on the ground. When we have the plan in place, and anything should happen, and an emergency occurs, we must execute our plan.

The take-off briefing rehearses the procedures in case anything happens after take-off:

> We fly straight ahead or whatever the plate tells. I always gesture the procedure, draw the emergency flight path, and point the directions we need to fly if anything happens. It is easier to share my understanding of what we must do.

The captain explained, gesturing with his hands to me as if to illustrate how he uses his hands and body language to explain the route to the first officer. The first officer nods in agreement.

The captain and first officer confirm that there are no additional or pressing factors they need to address for the upcoming take-off. The local weather is clear, and the winds are calm. "However," the captain explains,

> when there are winds and cloud layers in this area, it is company procedure to brief an instrument approach for the return to the airport if necessary. I can say: "We will have a

> moderate left crosswind on take-off. If we need to return to the airport immediately, we will directly proceed to so-and-so and execute the so-and-so approach."

The captain concludes the departure briefing: "Are there any questions?" The first officer shook his head and replied, "All good. We are on the same page," the first officer responded. "Let`s go."

The departure brief usually takes about a minute to get through. However, it is not a competition to do it as fast as possible:

> To me, the real focus of the departure briefing is the mindset of it. I try to encourage myself and the co-pilot to think, "Today, this engine will fail," because today is the day it just might. And when it does, we are prepared and have a plan.

The captain informs me.

The first officer then takes out the checklist, which is situated on the top of the forward instrument panel. The crew goes through the pre-flight checklist before starting the engines. Meanwhile, in the cabin, the attendants are hectic. Flight attendants are busy preparing the cabin and passengers before take-off. Each of the cabin crew has their designated responsibilities. There is a certain procedure to prepare the cabin. Flight attendants must check the restrooms and lock them before pushback. This serves the dual purpose of ensuring no passengers are inside and checking for any unusual odours or signs of smoke, which could indicate a problem.[3] The attendants visually inspect every row in the flight to ensure that loose items are stowed away correctly to prevent obstacles or hazards that could impede passengers during an evacuation. Passengers seated near emergency exits receive specific instructions on when and how to open them:

> It is the responsibility of the flight crew that the passengers sitting near the emergency exits are comfortable with the role they have to play if an emergency should occur. If they are uncomfortable, we relocate them to another seat.

A flight attendant explains (Informant Lisa 2023).

It is vital to have a clear line of sight through the cabin windows. This is not only to observe outside conditions for evacuation purposes, but also for spotting any issues, such as ice on the wing. While pilots also monitor this, the flight crew serves as additional eyes in the cabin. "That is why we do not allow the plastic slides on the windows before take-off. All seat belts must be securely fastened, ensuring that passengers are correctly protected during take-off. Seats must be in the upright position and tray tables stowed," Lisa, the flight attendant explained (Interview with Lisa 2023).

Just before the pushback[4] from the gate, the flight crew conducts the safety briefing for passengers. Infant seat belts and life vests are distributed as required, and passengers are instructed on the actions to be taken in the event of a water landing. The cockpit crew ensures that all necessary systems are operational, and the cabin crew verifies that the flight is secure, with safety equipment in place, including systems for door jettison and slide deployment.[5] Once all these tasks are completed, the cabin crew reports to the pilots in the cockpit that the cabin is clear and ready for take-off.[6]

3.2 ACCORDING TO THE PROCEDURES

The airlines' SOP (Standard Operating Procedures) cover a lot of the sets of step-by-step instructions, sometimes detailed, written instructions of who does what and when during the flight. A procedure is a "to-do list" or a timed schedule of tasks to be done according to the stages of the flight. Additionally, a flight is divided into stages. The SOP for each stage contains several groups of checklists, which form a system of informational aids used to reduce failure by compensating for potential limits of human memory and attention and help ensure pilots to out the various tasks that must be conducted.

A flight consists of several sequences of these standard operating procedures. In each flight operation, it is critical to perform the tasks required according to the procedures given. They are defined in a standard operating manual. Within the sequence, the checklists must be completed in a strict procedure. That is, one task becomes appropriate or even possible only when some others have been completed. Some tasks, and even a sequence of tasks, must be completed and verified as completed by both pilots before the next task is conducted. Within any sequence, the pilots engage in continuous communication to verify and coordinate.

Pilots routinely engage one another in their activities during any stage of a flight. Ahead of the pre-flight sequence, the pilots gather to discuss the flight plan, weather conditions, and any specific considerations. They share information about the aircraft's systems, fuel levels, and cargo to ensure that both pilots share a common view of the state of the aircraft, the route, and the flight ahead. In the cockpit, they conduct a series of checks to verify that the aircraft is in proper working condition.

Pilots go through the aircraft's systems by checklists and verbally confirm them. Each step is checked and cross-checked to ensure that the systems are working and nothing is missed. If one pilot calls out, "parking brake on," the other visually confirms and responds, "parking brake on." During taxiing, pilots communicate with air traffic control (ATC) and each other. They use a standardized phraseology to respond to each other's intentions and actions. During the planning phase, it is decided which of the pilots will be doing the take-off. After taxiing the aircraft to the end of the runway ahead of take-off, the captain underlines who oversees and holds the flight controls during the take-off sequence. If it is the captain who is doing it, he/she will conform: "My control." The first officer then immediately responds: "Your controls." If it is the first officer in control during take-off, the captain will say, "Your control." The first officer then immediately responds, "My control," and physically puts his or her hand on the control, signalling nonverbally that he or she is taking over control of the aircraft.

The following dialogue involves Captain Ove and First Officer Per preparing for the take-off on our flight. In the same way, it takes place on all of the company's flights of all flights:

Pilot A: (Reading from the checklist) "Cockpit preparation checklist, please verify."
Pilot B: "Cockpit preparation checklist."
Pilot A: "Parking brake set?"

Pilot B: "Parking brake set."
Pilot A: "Flight controls?"
Pilot B: (Physically moving the flight controls) "Flight controls"
Pilot A: "Follow me on the rudder. Left – Centre – Right – Centre 2

Pilot B has his feet on the rudder controls. While pilot A is pushing the rudder for the centre to the left, back to centre and then to the right, Pilot B is holding his feet on the rudder pedals on his side. The rudder input from Pilot A can then be felt by Pilot B.

While doing so, both pilots look at the rudder control indicator on the instrument panel. They both confirm that when pushing the rudder from centre to the right, back to the centre, then from the centre to the left, and that the rudder instrument indicates the same movement, they confirm that the rudder is working as it is supposed to. They mutually confirm it, and then proceed to the next item on the checklist.

Pilot A: "Instruments and radios?"
Pilot B: (Verifying instrument readings and adjusting radios) "Instruments and radios checked and set."
Pilot A: "Fuel quantity and balance?"
Pilot B: (Checking fuel gauges) "Fuel quantity and balance checked."
Pilot A: "Emergency equipment?"
Pilot B: (Verifying the presence of emergency equipment) "Emergency equipment verified."
Pilot A: "Briefing?"
Pilot B: (Summarizing key points of the pre-flight briefing) "Briefing completed."
Pilot A: "Before start checklist?"
Pilot B: "Before start checklist"
Pilot B: (Proceeds to read the next set of items on the checklist, and the dialogue continues accordingly.)

(*Observation of flight, Oslo, Norway 2019)*

After finishing the flight deck set-up, the cockpit crew asked the tower for permission to push back and start their engines.[7] The captain is responsible for taxiing on the ground, having the primary legal responsibility for the safe operation of the aircraft.[8]

Taxiing the aircraft can be challenging because of the proximity of other aircraft and ground vehicles, especially at large airports.

> Today, we have no problems with the taxiing, but during heavy snowfall and when all aircraft must de-ice before taking off, this airport is in chaos. Controlled chaos, I might add. Some days in winter is a living nightmare.

The captain comments.

Arriving at the designated runway after an uneventful taxi phase, the crew line the aircraft up and wait at a holding point before they are permitted onto the runway. An aircraft is allowed onto the runway only with the explicit permission of air traffic control. "A lesson learned," the first officer remarks when the tower asks us to "line up and wait at the holding point."[9] He is referring to the accident at Tenerife airport in 1977, in which two Boeing 747s collided on a fog-shrouded runway, claiming the lives of 583 people. To date, the world's worst non-terrorism-related air disaster.

After another aircraft lands, we are finally allowed onto the runway. The pilots line the airplane up on the runway centreline and wait for clearance to take off. After receiving the "cleared for take-off" instruction from the tower,[10] the two pilots look at each other. "Ready?" asked the captain. "Ready," the first officer replies.

"We have identified the hazards of the take-off and have mitigations in place," the captain continues. "We expect the take-off to go as planned - and we have set the boundaries for the departure route and will be using our normal departure procedures. In case of any emergencies, we use our company procedures."

"Checked," answers the first officer. The captain toggles both Flight Director (FD) switches to the "on" position.[11] First, on his side, as the pilot is flying, then on the first officer's side. Both pilots confirm to one another that they have the correct settings in the altitude and command speed windows on the mode control panel (MCP).[12] This device controls the autopilot and some of the flight instruments. Before pushback, the captain had commanded flap setting: "Set flaps 5," and at the same time, he used the electric trim switch on the yoke, trimming the aircraft to the correct aerodynamic configuration for take-off.

Once the tower gives clearance to take off, the flight crew confirms it. For an airplane to take off, its wings need to generate more lift than the force of gravity pulling it down.[13] Lift is created by the movement of air over the wings, and this requires the airplane to reach a certain speed. The faster the airplane moves, the more air flows over the wings, producing enough lift to counteract gravity.

To begin this process, the captain pushes the thrust levers forward, increasing engine power, which propels the airplane down the runway. "Setting thrust," the captain calls out and engages the auto-throttle.[14] The engines produce the necessary thrust to accelerate the airplane to take-off speed. As the plane gains speed, the captain uses the rudder to keep it straight on the runway and makes small adjustments with the ailerons to control any side-to-side movement caused by wind.

As the airplane accelerates, the first officer calls "speed alive" to confirm that the instruments are showing the plane is gaining speed. At 80 knots (nautical miles per hour), the speed is checked again to ensure everything is progressing as expected.

Soon, the airplane reaches a critical speed called V1.[15]This is the point at which the decision to take off is made - the plane is moving fast enough that stopping on the runway is no longer safe, so take-off must proceed. Once the aircraft reaches the calculated speed of 142 knots, the so-called Vr[16] rotation speed, the pilot not flying calls "rotate." The captain instantly, but slowly and purposely, initiates a smooth, continuous rotation to the initial target attitude. The captain draws the handles carefully but firmly towards himself and lifts the nose of the aircraft. The aircraft's nose raises smoothly towards the pre-set take-off pitch and accelerates because of the thrust of the engine. The altitude indicator shows the aircraft is ascending. The first officer monitors the instruments, and after double-checking both the airspeed and the vertical speed indicators, calls out "positive rate," indicating that the aircraft has left the runway and is airborne.

For an airplane to climb, it needs to maintain a balance between lift, gravity, thrust, and drag (the resistance the airplane feels as it moves through the air).

The engines continue to provide thrust, helping the plane gain both speed and altitude. Wind can also play a role in take-off. Taking off into the wind is preferred because it increases the airflow over the wings at lower ground speeds, which helps the airplane generate lift more efficiently.

The captain then calls for "gear up," and the first officer reaches for the undercarriage handle and raises it to the "gear up" position. The characteristic sound of the gear in travel fills the cockpit.[17] Directly above the undercarriage handle is a small display that, after a few seconds, illuminates three green lights. The first officer announces, "gear is up," confirming successful retraction. When the gear is eventually fully retracted reducing drag and allowing the airplane to continue accelerating and climbing.

The captain nods. After a few seconds, the captain starts to trim the aircraft to maintain minimum back pressure or "neutral stick" on the control column. Simultaneously, monitoring the flight displays, the flight path, and the aircraft's rate of ascent.

At around 1,000 feet above the airport elevation, the captain reduces the pitch attitude of the aircraft to increase acceleration. As the aircraft begins to gain speed, the captain requests flaps be moved progressively until fully retracted. This is a careful choreography done to ensure that flap operating speeds are not exceeded.

After a radio conversation with the tower, the first officer redirects his attention to the captain. After confirming by a simple nod that both parties have understood the air traffic control confirmation of their heading, the captain issues a new command: "after take-off checklist" (sometimes referred to as the "departure checklist"). The first officer then takes out the checklist on top of the instrument panel and reads out loud the text. The after-take-off checklist consists of a series of prompts the pilots must follow to change the aircraft's configuration. By going through the items on the list and performing the actions they mandate, the aircraft is configured for the next phase of the flight.

When passing the so-called transition level, all references to the height of the aircraft will change from feet above sea level to flight level, which is altitude relative to a fixed barometric pressure - a "standard atmosphere." Pilots adjust their altimeter settings from local pressure to the standard setting of 1013 hPa. This means that the altimeter will not be influenced by the local weather-induced air pressure.

Passing 10,000[18] feet marks the end of the climb phase of the flight, and the cruise phase commences. The pilots make an altitude call-out, "Passing 10 000 feet." At this altitude, pilots typically perform various checks and procedures to ensure everything is functioning correctly.[19] For example, the captain reaches for a switch on the overhead panel and turns off the "fasten seatbelt" signs. An audible signal accompanies the action, indicating to the cabin crew that in-flight services may begin.

3.3 A PRELIMINARY INTERNAL PEP-TALK

Sitting in the jump seat, I was struck by how much lies behind a seemingly routine take-off. By now, I had spent days following Ove and Christine, first in the crew room, then in the cabin, and now in the cockpit. I had seen how their work was shaped by procedures, briefings, and routines. Yet during take-off, I noticed that the formal scaffolding was not the whole picture.

From the jump seat, the sequence unfolded with practised ease, but beneath the flow of words and checklists, something else was happening. The captain leaned slightly forward, eyes fixed yet relaxed, his hand steady on the thrust levers. The first officer sat upright, scanning his instruments, giving short replies, but also quick glances across the cockpit. Their movements were measured, yet there was more than mechanics here - a sense of mutual reading, of anticipation, of making sense together in real time.

When the callouts ended, silence did not mean nothing happened. The silence itself was part of the rhythm, filled with the hum of the engines, the faint crackle of the radio, the breath between them. A nod, a hand shifting lightly on the yoke, the smallest of smiles. These gestures were as much about safety as the spoken commands, binding their individual tasks into a shared act.

The climb gradually eased into cruise. Shoulders lowered, voices softened, and the clipped tone of departure gave way to something looser. They did not step out of their roles - the calls and cross-checks continued - but within that structure came fragments of ordinary life. A comment about the early traffic over Oslo. A low laugh at a memory from a previous flight. The cockpit held both precision and humanity at once.

Watching them, I realized that safety was not a fixed condition or a set of technical instructions. It was something lived in these exchanges, carried in tone, in breath, in posture. It was constantly enacted and reinforced, sometimes even reinterpreted, as the aircraft moved through the sky. I reminded myself. Watch closely, listen carefully, and do not settle for simple explanations.

3.4 THE CLOSING PROCEDURE OF THE FLIGHT

On passing the first waypoint in the flight plan, the first officer re-establishes contact with air traffic control services and reports altitude, speed, and position. After first identifying the flight number, the air traffic controller responds by repeating the flight number, indicating that he has identified the aircraft on his radar system. After reading back the information, the controller commands the aircraft to proceed to the next waypoint. Although the captain hears the same radio transmission, the first officer repeats the information from air traffic control services aloud to him. On this flight, the air traffic controller changed the route of the aircraft, allowing it to fly directly to a waypoint further along the route. The initial flight plan assumed a predefined, templated route, a so-called SID (Standard instrument departure), which follows a series of waypoints flying from Oslo northbound to Tromsø airport. The crew immediately reacted to the order, turned their attention to the Flight computer, and found the waypoint as ordered by the controller. After receiving a confirmation from the first officer, the captain executes the change in the flight control computer system, confirming that the changes made by the captain are updated on the flight control unit on the first officer's side of the cockpit automatically.

The air traffic controller has an overview of the airspace. At any time, it can order the aircraft to change its speed, heading, and altitude. The purpose is to make use of the airspace as efficiently as possible, allowing as many flights as possible to take place simultaneously without jeopardizing safety.

Cruise is never just sitting back. Even with the autopilot engaged, the pilots remain in their roles. The pilot flying keeps a steady instrument scan, monitoring the flight path and aircraft performance, while the pilot monitoring handles radio calls and cross-checks the systems. Every 15 minutes, position reports are made, altimeters are cross-checked, and fuel figures are compared against the plan. Both watch hydraulics, electrics, pressurization, and engine indications with the same attention as in climb. They also think ahead: which airports are suitable if something happens now? What terrain lies below if an emergency descent should be required? Whether weather reports hint at turbulence or icing further along the route? The cockpit is calmer, yet vigilance does not relax.

As we climbed further, the rhythm in the cockpit began to shift. The rush of take-off was behind us, and the tempo softened. The captain leaned a little back in his seat, one hand still resting on the controls, the other loose on his thigh. The first officer checked the screens in front of him, then glanced across with a half-smile. They had flown together many times before, and the sense of being colleagues was mixed with something more like friendship.

The hum of the engines filled the pauses between their words. Outside, the sky opened wider with each thousand feet gained. Inside, the cockpit felt steady, held by the balance of trust that comes from knowing the other well. It was not a break from procedure. It was more the opposite. Precisely because the tasks were familiar and the trust between them strong, there was room to speak as more than just captain and first officer.

Then, the chime sounded from the cabin. A clear note cutting through the quiet. The attendant was checking in, asking if the pilots would like coffee or food. The captain gave a small nod, reached to acknowledge, and with that the rhythm of the cockpit folded back into the larger rhythm of the aircraft: "Coffee would be great."

Meanwhile, in the cabin, the passengers settle in. At cruising altitude, the cabin hums with activity. Some are reading, others chatting quietly, and a few children peer out of the windows, eyes wide with curiosity. The cabin crew begins their rounds, offering drinks and snacks, communicating calmly with passengers, and reassuring them about the smoothness of the flight.

A young passenger in seat 12A presses the call button, signalling a cabin crew member, Lina. She approaches with a warm smile and leans in, asking, "How can I help you?"

The passenger, visibly anxious, mentions they've never flown before and are feeling uneasy about the altitude. Lina listens attentively, her years of experience helping her spot this nervous energy. "How are you doing? Would you like some water or perhaps tea?" she suggests, her tone calming and reassuring. The passenger nods appreciatively, admitting that some water would be nice.

As Lina heads towards the galley, she checks in with the other attendants, making sure all service routines are running smoothly.

There is also an undercurrent of ongoing communication between the cockpit and the cabin crew. Periodic checks are made to ensure everything is running smoothly. The cockpit is informed of any potential concerns in the cabin and any information that might affect the passengers. One important information that is passed on from the pilots to the crew is upcoming turbulence, which in turn is communicated to the passengers, especially if they have information about any passengers who have expressed fear of flying.

In the cockpit, a soft chime alerts the pilots. Lina has buzzed in, requesting entry to bring them the coffee. With a few clicks, the captain unlocks the cockpit door, and Lina steps in. There is a quick exchange. Coffee cups are handed over. A few light words are exchanged about the smoothness of the flight. The door closes behind her as she steps back into the cabin.

Returning to the anxious passenger, she hands them a bottle of water and asks, "How are you feeling now?" The passenger, visibly more relaxed, expresses their thanks, noting how the crew's attentiveness has eased their worries.

Moments like these, unnoticed by most of the passengers, weave into the fabric of the flight. Through soft words, subtle gestures, and methodical attention to detail, the crew crafts an atmosphere of reassurance, one interaction at a time.

Back in the cockpit, the pilots continue their checks and monitoring, their routine punctuated by occasional instructions from air traffic control. They sip their coffee, fully aware that while their focus is on navigation and safety, the cabin crew manages an equally important task: caring for the passengers and ensuring their comfort.

In the controlled environment of the cockpit, everything feels measured and precise - each instrument calibrated, each command followed. In the cabin, it's more organic: emotions and interactions flow, connecting the crew to the passengers. The seamless coordination between both spaces. The technical precision in the cockpit and the human touch in the cabin create an intricate balance that sustains the flight. The sky outside is vast and indifferent, but within the aircraft, it's these small acts of care and coordination that keep everything running smoothly.

However, despite the calm atmosphere, the pilots have separate tasks to accomplish while the autopilot manages the flight path and altitude on cruise. The pilot monitoring continues to check in with an air traffic controller every 15 minutes or so in cruise as they pass between the waypoints of control.

Approach briefings are usually completed prior to beginning descent. On this flight, soon after we reached cruise level, the captain informed me: "Normally, we create a well thought out plan together with a detailed explanation of how we plan on descending into and approaching the airport we are landing at." The pilots explained how they obtain the weather at their destination and calculate the aircraft's landing performance based on the wind, temperature, and other atmospheric conditions. If there are any extra considerations, such as missed approaches or complicated taxi operations, the crew will discuss them.

The descent began long before the nose of the aircraft tilted downward. Already on the cruise, the captain and first officer had spoken of the challenges that lay ahead. Tromsø was no ordinary destination. It was a place where weather and terrain pressed hard on every decision. The Arctic itself demands constant negotiation. "With winter follows challenges" the captain said. "There are many missed approaches, constant icing conditions, complex fuel planning and a constant need to redo the planning as the weather changes. We must always monitor the aircraft's performance because the weather sets the limits for what the aircraft can actually do."

The first officer added quietly, almost as if reminding himself, "Turbulence, wind shear, circling while the runway is swept. It is normal. But it still demands everything of us. Knowing winter operations and having local knowledge about how we fly in and out is crucial."

The tone in the cockpit shifted as the top of descent[20] approached. The captain began the approach briefing, not by reading from a checklist but by drawing on years of experience. "Today we are flying into Tromsø. In summer there is daylight almost around the clock. In winter it is dark all day. Snow showers are common, and runways may close for sweeping no matter what you planned before departure. If one airport is marginal, the rest in the region are often the same. You must expect an ice-covered runway with reduced braking action. The latitude of Tromsø combined with steep terrain makes it a perfect place for icing." His words were not dramatic. They were steady, spoken as if he was letting the environment itself into the cockpit so it could be weighed along with every other factor.

Landing from the south at Tromsø is complicated. When passing the mountain south of the airport, the minimum safe altitude is 5000 feet when ten nautical miles from the runway, according to the terminal approach procedures for Tromsø (see Figure 3.2).

The following charts illustrate this.[21]

As the captain went through his approach briefing, he came to the point in the brief where he went through the information on the approach map (plate). "In case of a missed approach and go-around," he halted as he noticed my pen scrabbling on my notes as if he sensed that I wrote the word "go-around."[22] After finishing the briefing with the first officer, the captain turned to me with a more reflective tone. "The go-around is trained as one of the most normal things we do," he said. "It is simply what we do if the approach is not right. We climb away, reposition, and try again. The go-around procedure is there for safety. If the aircraft is not stable, we go around. No discussion." He paused for a moment, looking out through the windscreen as if rehearsing the situation in his mind.

> But still, many pilots carry a feeling that a go-around is a failure. It can sting your pride. You feel you should have made the landing work. Some even press on when they should not. That is when accidents happen.

I asked the captain how he thought about it. He leaned back slightly, his eyes still scanning the instruments, and said,

> The manual tells you when you must go around. But to be honest, no pilot really wants to. Some even see it as a kind of failure. We train for it every six months in the simulator, we know it by heart, but when it happens in real life, the cabin fills with tension. Passengers do not understand. They think something went wrong. You hear the sighs. Some grab their armrests.

He paused for a moment and glanced at the first officer, who nodded in agreement.

> But here is the thing. A go-around is not a failure. It is a decision. Sometimes the weather turns. Sometimes the approach is unstable. Sometimes another aircraft is still on the runway. The point is to act before the situation becomes unsafe. You have to take that decision without hesitation.

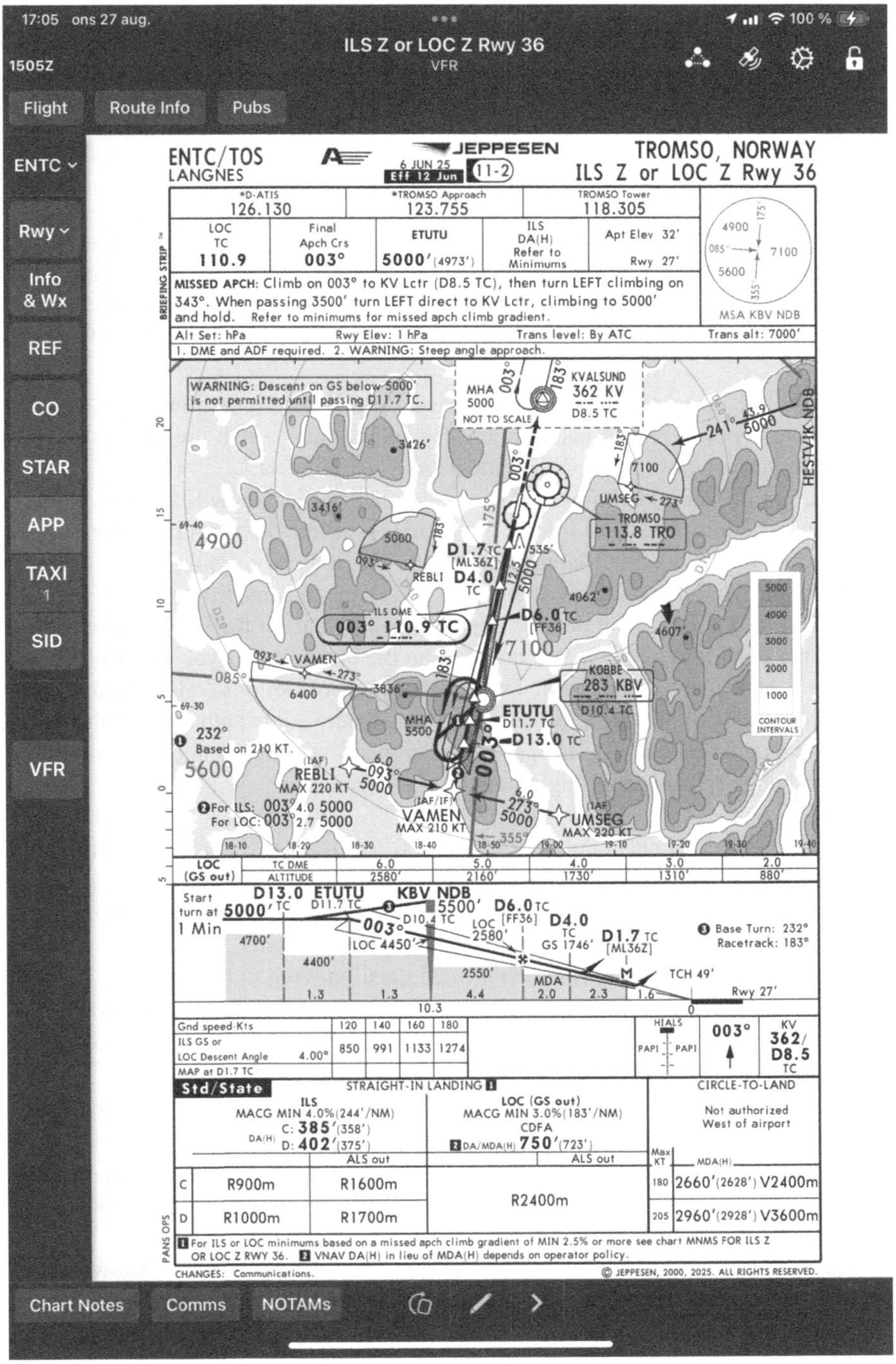

FIGURE 3.2 An approach plate is a standardized chart used under instrument flight rules to guide the aircraft from the en route structure to a safe landing. It contains all relevant navigation data for a specific approach. Fixes, bearings, altitudes, gradients, minima, and the defined

(Continued)

FIGURE 3.2 (Continued)
missed approach procedure. Frequencies for radio navigation aids and air traffic control are included, as are notes on lighting systems, runway elevation, airport layout, and obstacles in the vicinity. Each plate is tailored to a specific runway and approach type. In the picture, we see a precision approach to runway 36. The chart defines an initial fix, perhaps a VOR or RNAV waypoint, where the approach begins. From there, the aircraft descends along the published track, guided by the Instrument Landing System. Localizer frequency, inbound course, glide slope intercept altitude, and missed approach point are all clearly marked. The pilot follows a vertical and lateral path defined to ensure terrain clearance, separation from other traffic, and a stabilized approach profile. The minima will define the decision altitude and the point at which the pilot must either see the runway or execute a missed approach. The plate also outlines the missed approach path itself, including heading, climb gradient, and next fix.

He tapped the thrust levers lightly with his hand, almost absentmindedly. "It is better to give it power, climb back up and try again, than to risk pressing on with an approach you are not comfortable with. That is the mindset."

The first officer added quietly, "Passengers will never know that for us it is just another procedure. For them it feels dramatic. For us it is just part of flying."

The rest of the descent planning was done methodically. Using the "3 times the height" rule,[23] numbers were written, cross-checked, and entered into the flight computer. But the captain reminded me that this was the part of the flight where airmanship came most into play. "If you are high on the profile early, you still have time to recover. Later, there is no room. Then you are out of options." He traced with his hands how he would fly the approach, sketching the invisible path through the sky for his colleague to follow. The first officer responded with small nods, the rhythm of their exchange rehearsed through many flights together.

As the descent continued, the cockpit filled with the ritual of calls and confirmations.
"Set flaps fifteen," the captain commanded (after first have ordered flaps five which the first officer excecuted earlier on the descent).
"Flaps fifteen," the first officer confirmed, hand moving to the lever.
"Gear down" came a little later, the familiar thump of hydraulics following.
"Gear down, three green," replied the first officer, eyes flicking across the indicators.

Each action was read aloud, checked, and confirmed.

"When flying into Tromsø, we often put the gear down sooner," the captain explained. "It creates drag and steepens the descent path.[24] That way, we can get back to the glide slope."

Outside the cockpit windows, the Arctic coast came into view. Snow-covered ridges rose sharply from the sea. The approach path demanded precision because there was little room to manoeuvre if things went wrong. The pilots checked their altimeters against the latest pressure setting from air traffic control. Any discrepancy here could mean hundreds of feet of error close to the ground. "It is essential that the instruments are set correctly," the first officer said. "Everyone in congested airspace must be on the same setting. Otherwise, we do not know our true altitude."

Inside the aircraft, two worlds moved in parallel. In the cockpit, the descent was managed through checklists, briefings, and constant cross-checks. In the cabin, the attendants walked the aisles one final time. Seat belts fastened. Tray tables stowed. Toilets locked. Once everything was ready, the chief cabin attendant called the cockpit to confirm that the cabin was secure. Only then did the descent move into its final stage.

Tromsø normally offered an ILS approach. On this day, the autopilot locked onto the localizer and glide slope, guiding the aircraft steadily downward while both pilots monitored the instruments. The crew continuously scans the primary flying instruments and ensures that the correct airspeed is being flown. They monitor the localizer and glide-slope indications to ensure the ILS flight path is being maintained.[25]

Still, the captain reminded me that not all approaches were like this. "If the automatic systems cannot lock on, we fly what we call a two-dimensional approach.[26] That increases the workload a lot. Then the monitoring is intense."

As the ILS localizer and glide slope came alive, both pilots leaned into their scan.

The first officer announced "stabilised"[27] at the marker. This simple word carried weight. It meant that speed, configuration, and flight path were all within the strict limits required for a safe landing. Had the parameters not been within limits, the captain would have broken off and ordered a go-around.

At 1,000 feet,[28] the autopilot was disconnected. The captain flew by hand. His voice was calm. "Speed checked."

"Checked," the first officer replied.
"Landing checklist complete."
"Checklist complete"

From here on, it was all touch and feel. The captain held the nose steady, adjusting pitch with careful movements. The airspeed bled off, and the runway lights grew larger in the windshield. "Flare," he said softly as the aircraft floated a few feet above the ground. The thrust levers were pulled back to idle. The nose lifted slightly, and the back wheels kissed the runway.

"Speed brake up," the first officer called as the automatic system deployed.

For a few seconds, the aircraft rolled with its nose still high, then settled down onto the forward gear.

"Sixty knots." "Checked, my controls," the captain replied, taking the brakes.

"Oh, a kisser," the first officer remarked on the captain's smooth landing. "Well, we have a guest on the jump seat, so I had to show him my skills," the captain replied. They both laughed while keeping their eyes on the runway and being in the moment.

Reverse thrust faded as the speed dropped. The runway slid past the windows, snow piled high along its edges. The aircraft slowed to taxi pace and left the centreline, guided by instructions from the tower.

After landing, as the aircraft taxis along the tarmac towards its designated gate,[29] the shut-down sequence begins. This phase is a carefully choreographed series of actions that ensures the safe disengagement of the aircraft's systems and its passengers. The aircraft is expertly guided to its assigned parking area, or gate, under the instructions of air traffic control. With the aircraft securely positioned at the designated gate, the pilots initiate the orderly shutdown of the aircraft's engines. A thorough check is performed to secure all systems. Essential systems are deactivated. The lights are turned off.[30]

Inside the cabin, passengers exhaled, gathering their bags, reaching for coats. For most of them, the landing was already behind them. For the pilots, it was not finished until the after-landing scan was complete, frequencies changed, and taxi clearance obtained. The aircraft rolled towards the gate, guided into place by ground staff.

Only when the parking brake was set and the engines spooled down did the rhythm ease. Systems were switched off one by one. Lights dimmed. The hum of the aircraft faded to silence.

Passengers disembarked in steady rows, the cabin crew offering a final word or smile. On the flight deck, the debriefing began - not formal, but focused. How had the approach gone? What had the conditions demanded? What could be carried forward to the next flight? Safety is never a single act. It is an ongoing practice, one that continued even after the aircraft was at rest.

NOTES

1 The terms "flight deck" and "cockpit" are often used interchangeably but have slightly different connotations. The "flight deck" typically refers to the entire area where the flight crew operates, including pilot seats, control panels, and systems used for navigation and communication. It is more commonly used in formal or technical contexts, particularly in commercial and military aviation. The "cockpit," on the other hand, is a term more familiar to the general public and is often associated with smaller aircraft. It refers specifically to the area where the pilots sit and operate the controls, though in larger aircraft, this space can extend to include various control systems beyond the immediate pilot stations.

2 The pedestal in aviation refers to the instrument console located between the captain's and first officer's seats in the cockpit. It typically houses various controls and displays, such as communication systems, throttle levers, and navigational instruments. This area, along with the front and overhead panels, forms a key part of the cockpit's instrumentation layout.

3 The first indication of a fire is often smell.

4 A pushback is when the aircraft is moved away from the gate by a small tug vehicle. Large aircraft cannot reverse on their own, so this short backward movement positions them on the taxiway, ready to taxi under their own power.

5 Aircraft evacuation systems include mechanisms to jettison (or detach) doors in case of emergency, ensuring that they do not obstruct the exits during evacuation. Additionally, most commercial aircraft are equipped with inflatable evacuation slides, which automatically deploy when the doors are opened in an emergency mode. These slides ensure rapid and safe egress from the aircraft in emergency situations, such as an evacuation on the ground or during a water landing.

6 In aviation, a "secure cabin" refers to the state in which the cabin crew has completed all necessary safety checks to ensure the aircraft is ready for take-off or landing. This includes confirming that passengers are seated with seat belts fastened, tray tables stowed, seats in the upright position, and all carry-on luggage properly secured. Additionally, the cabin crew verifies that safety equipment, such as life vests, oxygen masks, and evacuation

slides, is in place and operational. Emergency exits and doors are also checked to ensure they are properly armed for emergency use. A final walkthrough is performed to confirm there are no loose items or hazards in the cabin. Once these checks are complete, the cabin crew reports to the cockpit that the cabin is "secure".

7 The timing of engine start varies depending on the aircraft type and operational procedures. For most aircraft, engine start occurs after pushback is initiated, once the aircraft has disconnected from ground power. However, in some cases or with certain aircraft, engines may be started before or during pushback. Ground power is typically used to provide electrical power to the aircraft while it is on the ground, and disconnecting from it is a necessary step before engine start if the aircraft's auxiliary power unit (APU) is not in use.

8 Many aircraft are equipped with a single "tiller" used for steering on the ground, which is installed on the captain's side of the flight deck. Some aircraft have tillers for both pilots, but it is up to the airline whether the first officer can taxi the aircraft if this option is available. Even if the first officer (often referred to as P2) is controlling the aircraft during taxiing, the captain (P1) typically retains ultimate legal responsibility for the aircraft's safe operation. In the event of a collision or incident, liability may still rest with the captain, as they are considered responsible for the flight's overall safety. This reflects a broader legal framework in aviation, where responsibility and accountability are often allocated in a way that defines legal blame rather than focusing solely on preventing accidents. Some industry professionals argue that legal structures in transport sectors, including aviation, tend to focus more on determining fault after incidents rather than proactively ensuring safety through shared responsibility.

9 The terms "holding point" and "line up and wait" can cause ambiguity. A holding point refers to a position on the taxiway just before the aircraft enters the runway, where it waits for clearance to proceed. "Line up and wait" refers to when an aircraft is given permission by air traffic control to enter the runway and align itself with the runway centreline, but must still wait for take-off clearance. In this instance, the aircraft has likely already accessed the runway and is positioned for take-off, but is awaiting further instructions.

10 Air traffic control (ATC) functions are divided into different roles to ensure the safe and efficient flow of air traffic. The ground controller manages aircraft movements on the taxiways and non-active runways, ensuring that planes move safely between gates and the runway. The tower controller is responsible for managing aircraft on active runways and during take-offs and landings. After take-off, responsibility is handed over to departure control, which manages the aircraft's initial climb and route away from the airport. En route aircraft are managed by area controllers, and as planes approach their destination, approach control takes over to guide them safely towards landing, at which point the aircraft transitions back to the tower and ground controllers at the destination airport.

11 The Flight Director (FD) is an essential component of an aircraft's autopilot system. It provides guidance to the pilots by displaying visual cues on the primary flight display (PFD), showing the recommended pitch and bank angles to maintain the desired flight path. While the FD can operate independently of the autopilot, it assists pilots in manually flying the aircraft more precisely by indicating how to adjust the aircraft's attitude to follow a pre-programmed flight route or specific manoeuvre.

12 The Mode Control Panel (MCP) is a key interface in modern aircraft used to control the autopilot and flight management system. Located in the cockpit, it allows pilots to set and adjust various flight parameters such as altitude, speed, heading, and vertical speed. By manipulating the MCP, pilots can engage or modify different modes of the autopilot to ensure the aircraft follows the desired flight path or adheres to air traffic control instructions during different phases of flight.

13 For an airplane to climb, the lift generated by the airplane's wings and airframe must be greater than the force gravity exerts on it.

14 Auto-throttle is a system used in many modern aircraft to automatically control engine power settings. Once activated by the pilot, it adjusts the thrust based on pre-programmed speed or performance targets, helping maintain optimal speed during different phases of flight (such as take-off, climb, and cruise). This reduces pilot workload and enhances efficiency by ensuring that thrust is adjusted in response to changing flight conditions, like altitude or speed. Auto-throttle is typically integrated with other systems like the autopilot, allowing for smooth and coordinated flight control.

15 V1 is the maximum speed that an aircraft can reach on the runway if an aborted take-off is to be affected safely. V1 is also the minimum speed at which the flight may proceed following a single-engine failure. It is determined for each aircraft through careful calculations and testing. Proper application of the V1 speed is crucial to ensuring the safety of the aircraft and passengers.

16 Vr, or rotation speed, is the speed at which the pilot lifts the aircraft's nose to achieve take-off. It is calculated based on several factors, including the aircraft's weight, flap settings, runway length, airfield elevation, temperature, wind conditions, and aircraft configuration. Heavier weights, higher temperatures, and higher altitudes typically result in a higher Vr. Pilots determine Vr using onboard flight management systems or performance charts, which ensure the aircraft reaches the optimal speed for safe lift-off while accounting for these variables.

17 The sound of the landing gear retracting can vary depending on the aircraft type. Typically, it is a combination of mechanical and hydraulic noises. This may include a low grinding or rumbling sound as the gear leaves its locked position, followed by an electric or hydraulic whine as the gear is pulled into the fuselage. The sound is often like that of the flaps being extended or retracted, but distinct to the system operating the landing gear.

18 Specific procedures may vary between airlines and regions, but 10,000 feet is a key point in flight. At this altitude, pilots may change their altimeter settings from the local pressure (QNH) to the standard setting of 1013.25 hPa to ensure uniform altitude references. Additionally, the United States enforces the "Sterile Flight Deck" rule below 10,000 feet, where only essential activities related to the flight are allowed to minimize distractions during critical phases of flight (U.S. FAR 121.542/135.100).

19 This phase of flight, though routine, illustrates the importance of rigorous procedures and attention to detail. The tragic Helios Airways Flight 522 crash in 2005 serves as a stark reminder of what can happen when such steps are overlooked. In that incident, the flight crew failed to properly set the cabin pressurization system, leaving it in manual mode instead of automatic. As the aircraft climbed, the cabin lost pressure, and the crew, unaware of the severity of the situation, failed to respond to multiple warnings. Eventually, the crew succumbed to hypoxia (lack of oxygen), and the plane continued autopilot until it ran out of fuel and crashed. This disaster underscores the importance of following protocols, particularly at critical stages like passing 10,000 feet.

20 Top of descent (TOD) is the calculated point where an aircraft begins its descent from cruise altitude, usually determined by the required vertical profile for the approach.

21 https://ww2.jeppesen.com/wp-content/uploads/2022/10/ENTC-rev-4-NOV-22.pdf, retrieved 11 oct 2024.

22 To reduce the likelihood of an unsuccessful landing due to the aircraft not being on the correct flight path or outside of reasonable flight parameters, in most airlines it is company policy and standard procedure that at a specified point on the final approach, if the aircraft is not within the stabilized approach criteria, a go-around must be initiated. A go-around is the procedure where a landing approach is discontinued, and the aircraft is flown back into climb. It can be initiated either by the crew or by air traffic control and is not considered an emergency. The decision is made if the approach is unstable, if there is traffic or obstacles on the runway, or if weather or wind conditions make a

safe landing uncertain. Executing a go-around involves applying full thrust, increasing pitch to a climb attitude, retracting the landing gear and flaps in stages, and following the missed approach procedure published for the airport.

23 The "3 times the height" rule is a general guideline used by pilots to estimate the required distance to descend. For example, if the aircraft is at 5,000 feet, multiplying this by three suggests that approximately 15 nautical miles are needed to reach the runway smoothly at an ideal descent rate. However, in certain airports – such as those in challenging environments like Northern Norway – factors like terrain, wind, and required speed reductions can complicate the descent, making it difficult to adhere to this simple rule.

When the aircraft is too high on the descent profile, the speed brake (a device on the wings that increases drag) is often deployed to help increase the descent rate without accelerating excessively, allowing the aircraft to regain the proper approach profile.

24 Lowering the undercarriage (gear) early is a standard technique in steep or difficult descents. It creates additional aerodynamic drag, helping the aircraft lose altitude without gaining excessive speed.

25 The *localizer* is part of the Instrument Landing System (ILS) that provides horizontal guidance to pilots, helping them align the aircraft with the runway centreline during the approach. The glideslope is another component of the ILS, offering vertical guidance to help pilots maintain the correct descent angle to the runway. Together, these indicators ensure that the aircraft follows the precise approach path during landing, minimizing the chances of deviation from the runway alignment or descent profile.

26 In aviation, there are two fundamental approach modes pilots use for safe landings, particularly when facing challenging weather conditions. A "2D approach" is principally navigating in two dimensions. The pilot relies on horizontal guidance from navigation aids and visual cues to line up the aircraft with the intended runway. However, these approaches lack precise vertical guidance to achieve an appropriate glide slope. Pilots must estimate the correct descent rate and altitude without vertical cues. It is not as precise as a complete 3D approach, providing horizontal and vertical guidance. A "non-precision approach" is a category of instrument approach without exact vertical guidance. These rely primarily on ground-based navigation aids including Very high frequency Omni-Range Radio (VOR) beacons, Non-Directional Beacons (NDB), or Global Positioning System (GPS) to steer the aircraft laterally along the correct path to the runway. During a non-precision approach, the pilot uses information about altitude and descent rate provided by the altimeter and visual cues including the distance from the runway threshold. Non-precision approaches are used when an airport lacks the equipment for precision approaches with glide slope guidance, such as an ILS (Instrument Landing System). It is important to note that non-precision approaches demand a higher level of skill and attention from pilots. They rely more on visual cues and pilot calculations. By contrast, precision approach systems such as ILS provide horizontal and vertical guidance, making landings more accurate and straightforward, especially when visibility is limited. In practice, pilots choose their approach based on the equipment available and the conditions at the destination airport. When precision approaches are an option, they are usually preferred because they offer greater safety and accuracy during landing.

27 The call *"stabilised"* is a standard confirmation made by the pilot monitoring during final approach, typically at or before 1000 feet above the runway elevation in instrument conditions (500 feet in visual conditions, depending on airline SOP). It signifies that the aircraft is within the operator's stabilized approach criteria: correct speed, correct configuration (flaps and gear set), correct power setting, and on the correct lateral and vertical path. If these conditions are not met, a go-around must be initiated. The stabilized call is therefore not merely a formality, but a safety gate designed to ensure that no unstable approaches are continued to after landing.

28 1000 feet under IMC conditions, or 500 feet if flying VMC.

29 While air traffic control (ATC) manages the aircraft's overall movement from touchdown to taxiing on the runway, the final guidance to the parking gate is typically handled by ground control. In some larges airports you can still see the use of a "Follow-Me" vehicle or an Airport Marshaller. The marshaller provides visual signals to guide the aircraft safely into its designated parking spot. In larger airports, a "Follow-Me" vehicle may assist in guiding the aircraft, especially in complex or busy environments, whereas in smaller airports, the marshaller may directly guide the aircraft without the use of such a vehicle.

30 The phrase "lights are turned off" refers to the deactivation of external lights, such as navigation and landing lights, which are no longer needed once the aircraft is parked. However, some internal lighting and non-essential systems are left powered until ground power is connected. Ground power supplied by the airport's ground support team, takes over from the aircraft's auxiliary power unit (APU) to maintain essential systems for the aircraft's preparation for its next flight, such as air conditioning, avionics, and lighting for cleaning or maintenance crews. Non-essential systems refer to those not critical for this phase, like certain in-flight entertainment or additional lighting.

4 Flight Safety

The Latin word *salus* means a state of being in which we are protected from harm or danger, feel secure, and are well-being. It captures what was sought long before "safety" became a technical term. That is the broad meaning of "safety," as it is also applied in aviation and flying operations. In aviation, safety is not a separate concern to be weighed against efficiency or performance. It is the foundation on which both rest. Without safety, there is no flight worth taking. For those who work in the field, safety is not a slogan pinned to a noticeboard. It has become a mindset in the industry. Safety measures are designed to prevent accidents and minimize the risk of injury or harm to people, property, and the environment.

Safety is a lived practice, embedded in each decision, checklist, and conversation that takes place before, during, and after a flight. It can be found in the briefest of exchanges: a glance between two pilots before take-off, the pause before acknowledging a clearance, or a slight change in tone when a situation demands heightened attention. During my fieldwork, I observed a crew preparing for departure from a busy airport. The flight was delayed, and the aircraft had been parked at a remote stand. The ground crew signalled that loading was complete, and the cargo door had been closed. Still, in the cockpit, the first officer noticed that the cargo door warning light had not extinguished. It was not the first time they had seen something like this - a faulty sensor, maybe, or a signal that just needed a moment to reset. But the captain did not brush it off. He asked for a pause in pushback and called the ground crew to double-check the latch. Not because procedures demanded it, but because experience had taught them that when something does not feel quite right, it rarely pays to ignore it. The delay was minor. But when the ground crew returned, he confirmed that the latch had not been properly engaged. The door was closed and looked secure from the outside, but it had not been locked into place. It was a small moment. A quiet choice. But it said everything about how safety works when it matters most. An experienced captain once told me, "Safety isn't something you switch on when you close the cockpit door. It's the way you think about every single part of the job. Even the parts that don't feel dangerous at the time." Another pilot put it more bluntly: "Safety is not a theoretical thing. It is something we deal with and do every moment" (Interview with John, 2023).

This view aligns with what safety science calls *work-as-done* – the reality of operational practice - as opposed to *work-as-imagined*, the version found in manuals and procedures (Clay-Williams, Austin, Braithwaite, & Hollnagel, 2021).[1] The phrase "to fly safe" is often heard. It means following the rules, checklists, and procedures. But it also means something more. It is about the mindset that the pilots and the crew bring into the cockpit and cabin. The ability to stay alert, even when things seem routine. The belief is that safety is worth slowing down for, even when schedules are tight. Safety is enacted through concrete actions. In the cross-checks, the challenge-and-response calls, the willingness to speak up and to notice the small things before

DOI: 10.1201/9781003725862-4

they become big. And the readiness to avoid even the smallest mistakes that sometimes have huge consequences.

Compared to other transportation forms, flying is safe. The aviation safety record is both a cultural and a practical achievement. It is cultural in that it depends on shared values. Safety is a core value in the professional culture among aviation professionals. It is in the creation of a safety culture that avoids risks and accidents and promotes a secure flying community. Sometimes, this highly skilled operational activity is a source of conflicts with the management in the flight organization.

Aviation safety management systems (SMS) aim to bridge that gap, but the bridge is never perfect. From a safety perspective, procedures are necessary. They provide structure and a common language. Yet safety is sustained in the countless acts of judgement that happen in the grey areas that those procedures cannot fully capture.

4.1 SKETCHING SAFETY HISTORY

The roots of aviation safety reach further back than the first powered flights. The very word aviation comes from the Latin *avis*, bird, with the suffix introduced by the French pioneer Guillaume de La Landelle in 1863. The wish to imitate birds inspired both the dream and the effort to fly. From the moment people looked up and imagined leaving the ground, the question of survival was there.

Early attempts often ended in failure, sometimes in death, but always in lessons. The story of Icarus is the oldest reminder of the notion of safety. His father Daedalus warned him not to fly too high or too low, but he ignored the advice, rose towards the sun, and fell to his death. The myth is a tale of ambition without limits. It tells us that courage is not enough, that safety depends on restraint, on knowing the margins. It is not a warning against the dream of flight, but against forgetting the limits that make it possible.

The first experiments of attemts to fly made this clear. People leapt from towers with wings of wood and cloth. They had imagination and bravery, but little understanding of lift, stability, or control. Each crash was proof that courage on its own could not keep them in the air.

Only when observation and science began to guide ambition did flight become more than a dream. Leonardo da Vinci sketched ideas of how air flowed over a wing. The Montgolfier brothers lifted their balloon in 1783, testing whether humans could even survive at altitude. A century later, at Kitty Hawk, the Wright brothers solved how to control the machine. With control came speed. With speed came new hazards.

In these early years, safety was not written in manuals or enforced by regulations. It was learned in the hangar, on the field, and too often in the wreckage. Pilots, engineers, and mechanics built their knowledge step by step, sometimes at great cost. Each failure left behind a lesson. Together, they became the foundations of what we now call aviation safety.

The twentieth century reshaped aviation safety under the shadow of war. Conflict speeds up technology. It pushes machines, and it pushes people. In the First World War, aircraft moved from fragile experiments to weapons. That change forced quick progress in structure, propulsion, and navigation.

The Second World War brought even more. Aircraft became larger, faster, and more complex. With that came new hazards. In the military, higher risks were accepted as part of the job. At the same time, many of the principles that later shaped civil aviation were first tested there. Out of war came structured training, standard procedures, and the first ideas of crew coordination.

After the war, commercial aviation carried more than new machines. It also carried the belief that safety could be worked on in a systematic way. Pressurized cabins, de-icing systems, and better navigation reduced some risks, but introduced others.

The Comet accidents in the early 1950s made this brutally clear. Metal fatigue around the square windows tore the jets apart. Propeller aircraft had never failed like that. The investigation that followed was among the first to use forensic engineering in detail. It set a new standard for how accidents were studied. The lesson was simple but profound. Safety could not rest on pilot skill alone. It had to be built into the aircraft itself, through engineering and a scientific understanding of failure.

By the late 1970s, it was clear that technology alone would not keep aviation safe. The real risk often lay in the cockpit. In how people spoke, listened, and acted together. Small misunderstandings, unspoken assumptions, or a first officer who chose silence could be as dangerous as any mechanical failure.

During my fieldwork, a senior captain told me how it had been. As a young first officer in the 1970s, he had lived through the shift: "When I started out, the cockpit was built on hierarchy. The captain spoke, the others listened. That was it. You didn't question, you didn't interrupt" (Interview with Helge 2004).

This recognition marked a turning point. Training moved away from safety as an individual skill and towards safety as a shared responsibility. Crew Resource Management[2] was introduced. It focused on communication, shared decision-making, and the courage to speak up even to a captain. At first, many dismissed it as "soft skills." But over time, it changed cockpit culture. Today, the ability to challenge authority respectfully is no luxury. In most Western airlines, it is seen as essential for safe flight.

The move to glass cockpits in the 1980s brought a new challenge. Digital displays gave pilots more information than ever before. They also gave more than could be handled. Automation sometimes behaved in ways no one expected. Crews found themselves asking: "What is the aircraft doing? And why is it doing that now?" (Interview Helge 2004).

It became clear that technology alone would not guarantee safety. Screens and systems could support, but they could also confuse. Pilots still needed the basics: to fly the aircraft, to talk, and to solve problems when the unexpected happens.

Every accident leaves a trace. Rules change, training is adjusted, procedures are rewritten. What happens in one part of the world is soon known everywhere else. Aviation is global, and so are its lessons. ICAO[3] has been central in this work, turning experience into standards that every state is expected to follow, even if each adapts them to its own context. Safety depends on people making sense of complex, dynamic situations and finding ways to hold them within the limits of what can be controlled.

The pattern is clear. Safety is never finished. It shifts and moves, always a step ahead or a step behind events. The cycle repeats itself: innovation, failure, investigation, and adaptation. The tools have become more advanced - black boxes, safety management systems, and modern simulators. But the essence is the same. Safety depends on people making sense of complex, dynamic situations and finding ways to hold them within the limits of what can be controlled.

4.2 FROM SAFETY SCIENCE TO SAFETY

Safety has never been easy to pin down. At its core lies a paradox. Safety is most visible when it is absent. Karl Weick once described reliability as a dynamic non-event, something that exists when nothing happens (Weick 2001). Hollnagel (2014) sharpened this by asking whether safety can ever be the object of science, or whether we are always studying its shadow - the accidents and failures that mark its absence.

Early attempts to explain accidents followed a simple logic. Heinrich's domino theory imagined accidents as linear sequences, where removing one cause would break the chain (Heinrich 1929). Later, Reason's Swiss Cheese Model offered a richer picture with several layers of defence and failure, though still rooted in cause and effect (Reason 1997). It marked a shift away from individual error and towards systemic conditions, challenging the old "bad apple" view that sought to identify and remove the guilty person. Reason's work reflected a broader turn in safety science, viewing accidents as the result of complex interactions between people, technology, and organization.

But this model, too, has limits. Sidney Dekker (2014) warned that the tidy alignment of holes in Reason`s model could create an illusion of linearity, as if accidents always followed a predictable path. In reality, failure often emerges from messy, non-linear patterns shaped by ongoing adaptation and trade-offs in everyday work.

Erik Hollnagel's distinction between Safety I and Safety II arose partly as a response to these shortcomings. Safety I focuses on preventing things from going wrong. Its methods involve identifying hazards, blocking deviations, and learning from adverse events. Safety II, on the other hand, is about ensuring that as many things as possible go right. It shifts the lens towards the adaptations and subtle shifts that make success possible under varying conditions.

Hollnagel (2014) has called this the causality credo. Accidents happen because something goes wrong. Causes can be found. Eliminating them will prevent recurrence (Hollnagel 2014; Schröder-Hinrichs et al. 2012). The credo is appealing, comforting, and reassuring.

An alternative approach to causality would be to challenge the assumption of causal agency itself. For centuries, the Western world has explained accidents in terms of cause and effect. This tradition is so accustomed to us that we hardly notice it. We cling to it, even as it becomes harder to reconcile with reality. Ludwig Wittgenstein captured the fragility of this belief in his "Tractatus Logico-Philosophicus" when he wrote that the events of the future cannot be inferred from those of the present. Superstition, he said, is the belief in the causal nexus (Wittgenstein 1922).

Despite the conceptual doubts, professionals are still drawn to these frameworks. Many safety improvements have grown out of precisely this kind of thinking. Each time the system faced a new form of failure, it responded by naming it: metal fatigue, human error, procedural violations, organizational breakdown, and safety culture.

And when the event seemed entirely without precedent, it was labelled a black swan (Balcerzak, Kostur, and Rajchel 2024).

As technical and socio-technical systems became more complex, older models struggled to keep up. The rise of information technology and automation made work less tractable and more dynamic (Hollnagel 2010; Perrow 1984). Tools that once seemed reliable became blunt instruments in a world no longer predictable. Safety managers, investigators, and regulators found themselves out of step with the work they were trying to understand.

This is the background for Hollnagel's formulation of Safety I and Safety II. Safety I defines safety as the absence of adverse outcomes. The aim is to prevent failure, stop the dominoes from falling. It aligns with a world imagined as linear and predictable. This has led to a wealth of useful practices, including accident investigation, root cause analysis, and barrier management. It remains deeply embedded in aviation training and regulation.

During fieldwork, I witnessed this logic in action. On a short domestic flight, the first officer noticed a mismatch between the load sheet and the fuel slip. The discrepancy was within tolerance, but the captain stopped the checklist and called for a recheck. "Better to stop it here than to carry it with us," he said (Observation 2013). This was Safety I at work - track the anomaly, resolve the cause, move on safely.

Safety II proposes something else. Safety is not the absence of failure, but the presence of adaptive capacity. It concerns the crew's ability to succeed under both expected and unexpected conditions (Hollnagel 2013, 2014). It invites us to study not just what goes wrong, but how people keep things going right. In practice, this means paying attention to small adjustments, improvisations, and shared awareness that hold the system together. In this way, Safety II overlaps with the field of resilience engineering (Hollnagel et al. 2010).

I also observed this approach. On a winter morning, a domestic crew faced de-icing delays, a congested apron, and a runway change minutes before departure. None of this was in the script. What made the operation successful was not a checklist, but the crew's ability to reorganize, reprioritize, and communicate clearly under time pressure. They kept the situation safe by staying flexible and composed.

Still, Safety II has not resolved the paradox at the heart of safety science. This is why some pilots view Safety II with scepticism. Not because they deny the value of adaptability or learning from success, but because the concept often feels detached from the real pressures of operational life. Hopeless, some say, because it sidelines the role of error, correction, and accountability. Helpless, because it offers little guidance when something actually goes wrong. In the cockpit, uncertainty cannot always be reframed as resilience. Decisions are made under time pressure, with consequences that are both immediate and irreversible. When things go wrong, as they sometimes do, pilots do not need abstract reflections on adaptive capacity. They need clarity, structure, and the kind of learning that is grounded in failure.

To speak of safety only in terms of what goes right, risks overlooking the very events that shape professional memory and drive improvement. It is often the moment when things nearly unravel that leaves the deepest impression. Not because of the outcome, but because of what it revealed about the system, the culture, or the crew. These are the events that fuel revisions, rewrites, and retellings.

It is simple logic. There can be no right-doing without wrong-doing. Removing wrongdoing from the safety narrative is to hollow it out. To speak of safety only in terms of success risks becoming blind to the very events that force learning and change.

Aviation makes visible the double logic that safety science cannot escape: there is no learning without both right-doing and wrong-doing. As a field, it has been both a donor and a beneficiary of safety science - contributing its errors, incidents, and hard-won lessons. It has offered its data, its stories to researchers. In return, it has borrowed theories, models, and methods to train its people, design its systems, and justify its rules.

The histories of safety science and CRM mirror each other in this respect. Both reflect an expanding understanding of where safety is located and how learning occurs. Both began with the individual - the pilot, the action, the moment. This was the cockpit level, where error and performance were seen av personal properties. Over time, attention shifted to the crew: communication, coordination, and the relational work of doing things together. Later still, safety was situated within the company and its wider conditions. This layered expansion resembles what Bronfenbrenner (1996) described in developmental psychology: an ecological movement from the micro to the meso and makro levels, where action can only be understood through its embedded contexts.

But the question remains. Have we defined safety, or only traced its edges? In expanding its analytical reach, has safety science clarified what safety is - or has it succeeded mainly in describing everything that surrounds it? The pilots and crew I have met point to a different understanding. Safety, in practice is not an object to be defined or secured, but a condition that must be continaually enacted - in how people meet, speak, and respond as situations shift. It is not what is controlles, but what is sustained through ongoing attunement, respect and readiness.

These observations are based on my ethnographic fieldwork with flight crews, where safety is encountered not as a concept, but as a practice. This enacted safety becomes visible in small, disciplined acts: calibrated speech, attentive listening, and the capacity to hold doubt without losing grip. Trust is built not through certainty, but through care. Perhaps this is the work - not to define safety once and for all, but to keep it in motion as a living ethic, renewed in every act of flight.

4.3 HAZARD, RISK, AND SAFETY

Hollnagel and others in the safety sciences have been concerned with the study of safety as a phenomenon. For those who work in aviation, the matter looks different. Safety is never experienced as an abstract principle. It shows itself through risk. Pilots and cabin crew rarely use the word safety when they speak about their work. What they talk about are the things that might go wrong and the ways to keep them from happening. A first officer once put it simply: "You never think about safety as such. You think about what might go wrong, and how to keep it from happening."

The safety science community has refined the terminology used for risk management. The word risk is a slippery one. The term risk is derived from the early Italian *risicare*, meaning "to dare" (Bernstein 1996a). The history of risk is the history of turning fate into choice. To fly has always been to dare, to push against natural limits, to make decisions in the face of uncertainty. In the literature, risk has been defined as the likelihood of loss, as uncertainty, or as the deviation between expected and actual outcomes (Vaughan and Elliott 1978). Other words are often tied to it, such as peril, hazard, and consequence. A peril is the source of loss, such as a storm, fire, or sabotage. A hazard is the condition that makes the loss more likely, such as turbulence, fatigue, or poor communication.

In aviation, many things that are otherwise harmless become hazardous when encountered in flight. A mountain is not in itself dangerous, but it becomes a hazard in icing winds or when a downdraft pushes an aircraft too close. Turbulence is only air in motion, but in the cabin, it can throw a passenger against the ceiling. Hazards also multiply. A minor fault may turn into a crisis when combined with a breakdown in communication. One captain told me as we flew across Northern Norway: "These mountains do not move. But the weather around them makes them unpredictable. Downdrafts, ice, turbulence. That is when they stop being scenery and turn into hazards."

There are formally four ways of handling risk. It can be avoided, reduced, shared, or accepted. Aviation uses all of them, a framework long recognized in risk science (Vaughan and Elliott 1978; Kolluru et al. 1996). Storms are avoided through rerouting. Pilot and crew fatigue is reduced through rules on rest and duty. Catastrophic but rare losses are shared through insurance. And some risks are accepted, carried by crews and passengers alike as the unavoidable price of flight.

The complete elimination of risk in aviation is impossible. Perfect safety would require every aircraft to be grounded and aviation itself to end. Some risks can only be managed by avoidance. Others can be reduced but never fully removed. Beyond that, they must be shared or tolerated. Different parts of aviation accept different levels of risk. Military flying operates with tolerances that would never be acceptable in civil aviation. Low-level flight at high speed may be routine for a fighter pilot, but it would never be permitted for passengers. Civil aviation does not normally define risk thresholds in numbers, but in practice, they are enforced through law, equipment standards, and absolute prohibitions. Alcohol in the cockpit is one example. It is not to be managed or mitigated but banned outright.

The principle of ALARP, as low as reasonably practicable, underpins much of this. Risks are reduced as far as possible, unless the effort of further reduction would be wholly out of proportion to the benefit. This is why airlines invest heavily in training, monitoring, and redundancy. Aviation is sustained not by removing risk but by living with it.

For passengers, aviation safety is experienced. You enter a cabin where the lights are dimmed, the crew moves without hurry, and the announcements are measured. You may not think about risk at all. The sense of safety comes not only from the physical environment but from the social cues around you. A calm voice on the intercom, the familiar act of fastening a seatbelt, the way other passengers settle in their seats. For passengers, aviation safety is not technical. It is social.

The experience of safety depends on trust. Trust that the aircraft is maintained and that the pilots are competent, that the system itself is watching. Passengers do not see the hazards that the crew manages, but they rely on the belief that they are contained. As one first officer told me, "What passengers experience as safety is really trust. They do not see the risks, they see the order we create around them" (Interview 2023). I once sat on a flight into Oslo where turbulence rattled the cabin so hard that trays and cups jumped. Conversations stopped. Some passengers held on to their armrests. What kept the situation calm was not the end of the turbulence but the way the crew carried themselves. The cabin attendant walked slowly down the aisle, paused to check on a passenger who had turned pale, and spoke quietly to her colleagues. When the captain came on the radio and said, "We expect this for another ten minutes, but it is well within limits," the cabin exhaled together. What might have been a moment of fear became a moment of reassurance. Safety was not only the strength of the aircraft but also the way trust was renewed through the crew's behaviour.

For pilots and crew, safety is never invisible. It is continuous work. Hazards appear in large and small forms. A snowstorm closing in on approach, a smoke detector that flickers once, or a passenger who does not answer when spoken to. Safety lies in catching these things early, in holding them within limits. Checklists and procedures provide a framework, but they never guarantee the outcome. The guarantee is created in the way people use them, in how they interpret silence, hesitation, or the feel of the aircraft through the controls.

On a winter evening, I joined a flight into Tromsø. Snow was falling heavily. The captain briefed simply: "We will try once. If the runway is too slippery, we divert." For the passengers, the cabin felt routine, even quiet. But the pilots in the cockpit spent every second weighing where to draw the line between acceptable and unacceptable risk. Safety in that moment was not the absence of danger but the containment of it. An experienced captain told me,

> If you have never been truly afraid, you do not really understand safety. The feeling of hanging on with your bare hands, knowing that if you let go you will get badly hurt, that is what teaches you why safety matters.
>
> *(Interview with Evert, 2024)*

For him, safety was not certainty, but recognition of uncertainty.

Some airlines insist on strict compliance with procedures. Others leave space for what happens between the written steps. One company I observed even encouraged manual flying on clear days without the instrument landing system, to keep skills alive. "Otherwise," a captain told me, "the day the system fails, you will not know how to handle it" (Interview Evert 2024). What looked like play was in fact preparation for failure.

This is also what risk science calls dynamic risk, the continuous assessment of risk in rapidly changing circumstances. Decisions under such conditions cannot be reduced to manuals. Pilots make sense of cues that may be visual, acoustic, or tactile, and build mental models of situations as they unfold. Sensitivity to these cues and the ability to act on them is not only a matter of training, but of lived experience.

Safety in aviation is not only created by the individuals in the cockpit. The system itself is open and shaped by forces outside. Governments, regulators, manufacturers, and international organizations such as ICAO,[4] EASA, and IATA all set boundaries. Economics and politics also shape how safety is managed. Geopolitical shifts can alter routes overnight. Sanctions may decide whether spare parts are available. Cultural norms affect how risk is understood, whether hierarchies in the cockpit are strict, or whether crews can speak more openly. All of this frames how safety is lived in practice. Lindbergh himself once wrote that part of the attraction of flight was the willingness to accept risk for the sheer joy of being airborne (Lindbergh 1927). That remains true, though it is now framed by far more complex layers of law, technology, and organization.

In 2006, ICAO required all member states to implement Safety Management Systems.[5] The purpose was to move risk management from an informal practice at the sharp end to a systematic responsibility for management. Companies now collect data on hazards, analyse risks, and make decisions in advance. This is often described as a proactive approach. But systems only work if the flow between reporting and action is trusted. Without trust, hazards remain hidden.

For passengers, safety appears as an order. For the crew, it is vigilance. For airlines and authorities, it is systems and procedures. In practice, these layers meet in small actions. A boarding that feels calm. A measured voice over the intercom. A captain who chooses to go around rather than force a landing.

Safety is not the absence of risk. It is the continuous work of holding risk within limits that can be controlled. It lives in the fragile space between hazard and consequence. That is where safety is created, flight by flight, moment by moment.

4.4 THE CREW AS SAFETY GUARDS

My first encounter with aviation was on Svalbard. I had been invited to visit the base of Lufttransport in Longyearbyen, where small aircraft and helicopters operated in some of the world's harshest conditions. On the door leading from the crew room into the hangar, there was a photograph. It showed a light aircraft that had crashed into a tree. Beneath the image was a sentence that struck me with lasting force: "Flying is not inherently dangerous. But to an even greater degree than the sea, it is terribly unforgiving of any carelessness, incapacity, or neglect." The phrase is widely attributed to Captain Alfred Gilmer Lamplugh of the British Aviation Insurance Group in the early 1930s. Over the years, the phrase has been reproduced on countless plaques and posters. What mattered to me was not its origin but its placement. On Svalbard, it hung as the last reminder to pilots before they took an aircraft out of the hangar to fly special missions on behalf of the Norwegian Coast Guard. It captured the essence of aviation safety in one simple message. Risk is ever-present. Neglect will not be forgiven. Whether the attribution is precise or not, the phrase has become a kind of folk wisdom in aviation communities. It conveys the simple truth that flying leaves little room for error (see Figure 4.1).

Safety is at the core of aviation. If safety breaks down, nothing else matters. The risks involved in flight can be extreme, and the consequences can be fatal. Yet, safety is not only the sum of what individual pilots or cabin attendants do. It is a shared responsibility woven into the system of procedures, routines, and regulations that shape daily flight operations. In practice, the crew becomes the visible guardians of safety.

Pilots and cabin attendants rarely talk about safety as an abstract principle. They speak of the things that make it possible. They talk about checklists, routines, cross-checks, and the small gestures of cooperation that prevent small deviations from becoming serious. Safety is in the way a captain briefs a first officer before descent. It is in the way a cabin attendant notices a passenger behaving strangely and informs the cockpit. It is in the way a deviation from a checklist is caught, not with blame, but with a quiet correction. Safety is not a slogan pinned on a wall. It is a practice lived through speech, gestures, and attention.

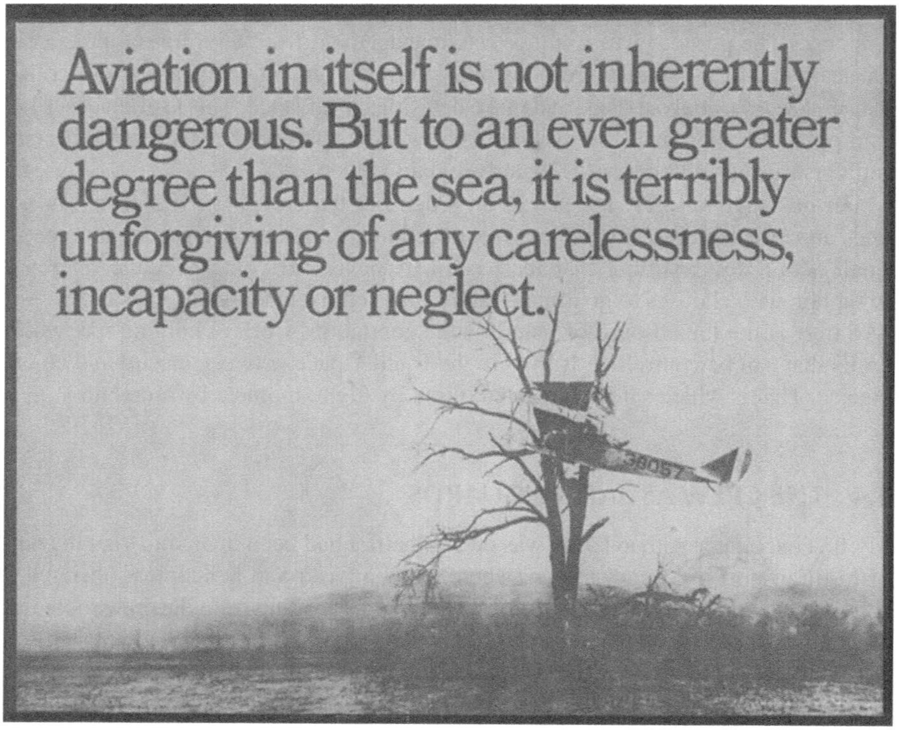

FIGURE 4.1 A photo of the Image I saw hanging on the door leading into the Hangar. Unknown source.

(Photo of the image: Vegard Nergård)

Reason (1997) and others have argued that aviation became the field in which the very idea of a safety culture was developed, before spreading to other industries. The selection, training, and recurrent checking of pilots and cabin crew, the ways in which crews handle checklists, manage cockpit communication, and work with standard operating procedures have been benchmarked and studied by medicine, nuclear power, and offshore engineering (Reason 1997).

Aviation is one of the most tightly coupled systems in existence. Once a flight is underway, processes move too quickly to allow pause or reversal. Perrow described such systems as prone to "normal accidents," because complexity and speed make errors inevitable (Perrow 1984). To guard against this, aviation has created standards and regulations that stretch from the design table to the hangar floor. This is why standards and procedures in aviation extend from the design table to the hangar floor, from the simulator to the cockpit.

Yet in spite of the technical systems and legal frameworks, safety always returns to the human level. Pilots and cabin attendants sometimes say that they are both the starting point and the end point of error. In the earliest years of safety science, Helmreich and Reason stressed the importance of seeing crews at "the sharp end" of safety, where decisions and perceptions matter most (Helmreich 1998). An experienced captain once remarked to me: "There are no tough old guys in aviation, for all of them died early in their quest to become tough" (Interview Hans 2004).

Safety is not a matter of strength but of humility and attention. This humility is made tangible in simulator training. Crews are placed into situations that in real life would be catastrophic. An engine fails on take-off. Smoke fills the cockpit. Hydraulic systems show contradictory warnings. These scenarios are not designed to trap the pilots but to rehearse coordination under pressure. In one simulator session, I observed that the instructor deliberately misread a checklist item, skipping a step. The crew caught the mistake, corrected it, and moved on. In the debrief afterwards, the instructor explained why he had done it. Errors creep into routine, he said. It is not about whether they occur, but whether you as a crew detect them before they cascade. Such sessions are demanding, but they train vigilance and communication in ways that no lecture could achieve. In the cabin crew training we find similar approaches. During one of the many recurrent trainings I have witnessed, attendants were asked to deal with impossible-seeming situations. A fire in a galley oven while turbulence makes standing almost impossible. An aggressive passenger, while another passenger collapses in a medical emergency. The point is not that these exact combinations will occur but that the crew learn to divide attention, delegate tasks, and keep communication flowing. The learning lies as much in the discussion afterwards as in the exercise itself. One attendant put it simply: "We do not learn alone, we learn together" (Interview with Lise, 2024).

This collective orientation is sustained by what aviation calls just culture. In this system, errors are reported without fear of automatic punishment. The emphasis is on understanding why the mistake happened, not only on assigning blame. If a pilot misses a radio call or a cabin attendant forgets to arm a door, the expectation is that the error is logged and analysed. Unless gross negligence or deliberate violation is involved, the individual is not punished. Instead, the organization learns. Just culture is fragile because it depends on trust. If crew members fear that reporting an error will harm their career, they will stay silent. If they are confident that their honesty will improve safety, they will report. This is one of aviation's key lessons for other sectors. Without trust, learning stalls.

At times, this openness extends into daily flight operations. On one short-haul flight I joined in northern Norway, the captain turned to his first officer after an unstable approach and said that we should have gone around sooner. His tone was calm. The first officer nodded. The event was logged after landing. There was no shouting, no humiliation, only a quiet recognition that they had allowed the approach to continue longer than company policy allowed. For the airline, the fact that this was reported was as important as the fact that the flight landed safely. Learning could only happen because the captain felt confident that admission of a lapse would not bring punishment.

The same orientation can be seen in search and rescue operations. Ash (2023) describes a mission in which a helicopter crew on Svalbard was tasked with evacuating a sick fisherman. Arriving on scene, they encountered dense fog and calm seas that disabled the Doppler system used for precision hovering. Rather than pressing on, they contacted another vessel, found the edge of the fog bank, and relocated the patient transfer. Later, they heard that another crew had solved a similar problem by using a vessel's wake to generate a Doppler return. Both solutions were shared. Neither was punished for deviating from the standard procedure. In fact, the improvisations were valued because they expanded the repertoire of safe options.

Modern aviation operates within an intricate web of international and national regulations. The world is divided into Flight Information Regions, each governed by a state authority. The International Civil Aviation Organization recommends standards and practices that states translate into law, and airlines are required to comply at all times. Compliance itself is a form of safety assurance, but it is not sufficient. An airline that wishes to be truly safe must cultivate what is often called a safety culture (ICAO 2006).

A safety culture is not built overnight. It grows through leadership, communication, and the willingness of employees to report mistakes without fear of reprisal. In such a culture, safety is always open to improvement. A safety culture develops in the daily interactions between captains and first officers, between senior and junior cabin crew, and between crews and ground staff. It is a learning culture, where incidents and near misses are analysed and shared, and where solutions circulate informally between crews as much as they are written into official reports. Engineers, too, are part of this web. They ensure that aircraft are maintained according to the highest standards, applying inspection and repair protocols with the same vigilance that flight crews apply to checklists. In conversations with maintenance staff, I was often struck by their awareness that a missed detail in the hangar could resurface as a crisis in the sky.

The risks of flight can be extreme, and the consequences can be fatal. Yet, most flights pass without incident. This is not because risks are absent, but because they are managed. Accident prevention involves procedures, training, and organizational routines that extend from cockpit to cabin to hangar. But it also involves the small, uncelebrated decisions. A flight attendant insists on rechecking a latch. A pilot diverts early rather than pressing into worsening weather. An engineer replaces a part a few hours ahead of schedule because it does not sound right. These acts rarely appear in statistics, but they are what keep passengers safe.

In this sense, crews stand as both vulnerable and strong. They are vulnerable because they operate in systems where small errors can be fatal. But they are strong because they create safety through vigilance, cooperation, and humility. They are guardians who embody safety in practice, turning rules and procedures into lived action.

4.5 THE AUTOMATION FALLACY

Aviation safety is intimately linked to automation. Over the past decades, new tools have entered the cockpit and changed the very practice of flying. Radar, GPS, and collision avoidance systems have given crews capabilities that earlier generations could only imagine. At the same time, automation has removed many of the repetitive and routine tasks that once distracted from the act of flying itself. Yet, the story of automation always circles back to the same tension. Technology has undoubtedly enhanced safety, but it has also introduced new risks when the human being is left to monitor complex systems that do not always behave as expected.

Several captains told me that the greatest danger is not when automation fails outright, but when it fails in ways that are opaque to the crew. A first officer in Gardermoen once explained it after a long simulator session. When everything goes as planned, automation is a gift. When the picture collapses, the systems can deepen the chaos. He pointed to the exercise they had just completed. In poor weather, the autopilot disconnected, and the aircraft entered an unusual attitude. Their task was to regain control without the help of instruments. Within minutes, they were sweating, not because the task was physically demanding but because the pressure to act correctly had arrived without warning. The simulator was designed for this kind of failure training. By being thrown into the unknown, they created a memory of how to improvise when systems offered no support. This logic of training reflects what Bainbridge (1983) once called the ironies of automation, where human beings are left with the hardest job of all: to take over in rare but critical situations.

The shadow of Air France flight 447 hung over many such discussions. When the Airbus A330 disappeared into the Atlantic in 2009, killing 228 people, the official investigation concluded that automatic systems had failed, but also that the crew were unable to regain control manually (BEA 2012). For many in aviation, the accident became a warning that automation can create an illusion of safety while quietly eroding the very skills required when safety collapses. Instructors often stress this point in simulator training. They remind their students that the machine can do most things better than we can, but only the pilot can land without power or without instruments.

The word deskilling is often heard in crew rooms. The more the systems take over, the more pilots risk losing their intuitive grasp. Research has described how extensive use of automation can lead to overreliance and the loss of psychomotor and cognitive skills (Damos et al. 1999, Hutchins et al. 1999, Sarter & Woods 1997). Parasuraman and Riley (1997) even called it a vicious cycle. As pilots recognize their own skill loss, they come to rely even more heavily on automation. Instructors in the Air Ambulance in Norway told me they tried to counteract this by encouraging first officers to land manually whenever the situation allowed it. One captain said he wanted his crew to keep alive the feeling of hand flying, because when automation falters, it is still the human being who must act.

Automation also brings a cultural question. Pilots often describe the joy of hand flying, yet they also know that the profession is moving steadily towards more machine-driven flight. Many younger colleagues engage the autopilot earlier and disconnect it later, not because they dislike flying but because they have grown up in a generation of aircraft designed to fly themselves. One captain described it as sitting in a chair that suddenly tips you off balance. You must know when to intervene, but you must also know what the system is doing. That understanding is not always straightforward.

This unease has been present since the early days of cockpit computerization. As Wiener observed in 1988, automation was introduced so quickly that pilots, designers, and operators struggled to formulate an overall strategy for its use. The human factors profession was, as he put it, struggling to catch up. Two decades later, Ferris, Sarter, and Wickens (2007) argued that many of the same issues remained unresolved. New ones had even emerged as the systems became more complex and more deeply embedded in daily operations.

Today, automation has reached a new frontier with the arrival of artificial intelligence. Some regulators and manufacturers speak openly of replacing the two-pilot concept with a single-pilot model in long-haul operations. Among the crews I have spoken with, this idea is met with scepticism. They argue that automation can reduce workload but not replace the shared judgement that emerges in dialogue between the captain and the first officer. A system can assist, but it cannot take responsibility.

These debates show why the term automation fallacy is used. It is a fallacy to believe that technology alone can make us safe. No matter how advanced the systems become, they cannot replace the attention, cooperation, and humility that crews bring to their work. Automation has transformed the practice of aviation, but the craft of flying remains a human responsibility.

NOTES

1 Clay-Williams, R., Austin, E., Braithwaite, J., Hollnagel, E., & Rapport, F. (2021). Qualitative assessment to improve everyday activities: Work-as-imagined and work-as-done. In *Transforming healthcare with qualitative research* (1st ed., pp. 71–82). Routledge. https://doi.org/10.4324/9780429299797-11.

2 Crew Resource Management (CRM) emerged in the late 1970s following a series of accidents where inadequate communication and poor teamwork were identified as contributing factors (Helmreich, 1999). It is now a regulatory requirement in commercial aviation, emphasizing non-technical skills such as communication, decision-making, situational awareness, and leadership, taught alongside technical training in recurrent courses.

3 The International Civil Aviation Organization (ICAO) is a specialized agency of the United Nations, established in 1944 under the Chicago Convention. ICAO sets international standards and regulations for aviation safety, security, efficiency, and environmental protection. It facilitates cooperation among its 193 member states to ensure a safe and orderly development of international civil aviation. By developing Standards and Recommended Practices (SARPs), ICAO provides a unified framework for global aviation operations, ensuring consistency in air travel across nations.

4 The International Civil Aviation Organization (ICAO) is a specialized agency of the United Nations, established in 1944 under the Chicago Convention. ICAO sets international standards and regulations for aviation safety, security, efficiency, and environmental protection. It facilitates cooperation among its 193 member states to ensure a safe and orderly development of international civil aviation. By developing Standards and Recommended Practices (SARPs), ICAO provides a unified framework for global aviation operations, ensuring consistency in air travel across nations.

5 The Safety Management System (SMS*)* in an airline is a structured framework designed to proactively manage safety risks. It comprises four main components: Safety Policy, which sets the airline's safety objectives, responsibilities, and organizational structure; Safety Risk Management (SRM), where hazards are identified, risks assessed, and mitigation strategies implemented; Safety Assurance, which evaluates the effectiveness of safety measures through monitoring, audits, and performance indicators; and Safety Promotion, which emphasizes training, communication, and fostering a culture of safety across the organization. In practice, SMS integrates technology, such as real-time data monitoring, with human elements, including collaborative reporting and regular training. This ensures that pilots, cabin crew, maintenance teams, and ground staff work together to identify and address safety concerns, creating a feedback loop for continuous improvement and resilience in operations.

5 Creating Safety

Daily life practices of flight crews are based on team interaction to ensure safety. The flight cockpit is a microcosmos dominated by technic that serves and is served by humans.

In the cockpit, the captain and the first officer share what they see, know, and think. Safety and efficiency are the soul of their doing, and they continuously articulate and report on it. They practice a sort of distributed cognition in a system of observing, acting, and communication. This extremely accurate coordination and communication is expressed in the cockpit jargon as "being on the same page." It is not just about sharing observations and information but also involves the coordinated use of tools and instruments. It also includes the physical environment and air traffic control. Pilots are taught from their first day in service that they never operate in sole isolation. Their capability to think and act together enables the system to master its tasks. To fly an aircraft safely depends on the coordination of multiple individual resources that form the competent skill to fly an aircraft safely.

5.1 BEING ON THE SAME PAGE

Pilots know the importance of observing themselves and each other. They share a common perception and being in the cockpit. They are trained to be on the same page, and they continuously need to agree on sharing understanding and opinion with one another. The captain and first officer need a clear common understanding of any situations that occur during the flight:

> When we are on the same page we are aligned in our thoughts, how we perceive the operation and that we have the same expectations about it. It is not that we are in a constant state of harmony but share the certainty of doing things correct. On the contrary, sometimes being on the same page involves disagreement but then we must cooperate and communicate on what we see and perceive to come to a mutual understanding of the nuances on what we are doing.
>
> *(interview with first officer "Ole," 2018)*

Being on the same page further means the pilots work strictly coordinated on the procedures settled to operate the aircraft. It involves in addition to the activity in the cockpit, communicating with air traffic control and simultaneously monitoring the various instruments and systems. The pilots are in continuous communication with the aircraft itself through its automation system, assisting navigation and communication. The air traffic controllers take an active part in the preparation of a flight to ensure that everyone involved in the operation is on track. All these resources are coordinated as one cognitive system.

"Being on the same page" is a statement about a mindset in flight operations. The thorough sharing of actions and opinions related to the tasks in the flight operations

DOI: 10.1201/9781003725862-5

contextualizes the individual cognitive workload. A flight operation relies on these coordinated efforts of the pilots working together. As in any dynamic setting, the cooperation between the pilots and the rest of the flight crew must be flexible and adaptable to changing circumstances. This calls for quick responses and adjustment of the "collective intelligence" shared by the crew.[1]

Tragic examples of failures occur if and when pilots on a flight deck do not stick to the pilot saying, "Two heads must operate as one." The pilot culture has a high recognition of the interdependence that pilots on the flight deck need to have with one another. My years of fieldwork talking with pilots have underlined the basics of all flight operations: no individual possesses all the knowledge and skills required for every situation. A pilot flying in the Norwegian Air Ambulance services claimed, when describing his crew colleagues:

> We are entirely dependent of one another, and everyone onboard must contribute. All in the crew are equally important. It is an absolute must to foster what I would call a collective intelligence. Individually, we might well be stupid, but together we are way better off. When we bring our heads together, we create an `intelligencè that surpasses individual limitations.
>
> *(Interview with Captain "Tom" 2019)*

All throughout the flight operation, the pilots work intensively together to ensure the operation. During a flight, they work in coordination on different systems on the aircraft. While dealing with different tasks simultaneously, the communication between them is of the utmost professional norm. They share opinions about their actions and give each other feedback and support. The successful and safe flight is a collective endeavour, in which each of the pilots' actions affects the overall outcome:

> It is not the communication systems we have onboard, nor are the words, all the terms we use. The essence of any flight operation is the safe outcome of it. "In the cockpit, two heads must become one," an experienced captain explained.
>
> *(Interview with Captain "Tom" 2019)*

Sometimes sharing one's perception of a situation may result in a disagreement. Pilots are trained to articulate their view and perception, which sometimes may challenge the cockpit companion in a given situation. This sharing also acknowledges different perceptions and promotes an ideal in the communication between pilots. The mode of sharing and challenging one another is fundamental in safe flying.

Flying an aircraft from one destination to another involves several participants not residing in the aircraft. Pilots collaborate with air traffic control services according to the planned flight route to check altitude changes, other aircraft, safe take-offs, landings, and other en-route operations. It is equally important that the pilots align with the air control services and other aircraft in the airspace. Pilots in the cockpit are on a moment-to-moment basis aware of what they are doing and why.[2]

Once airborne, pilots continually scan the aircraft's instruments, including the attitude indicator, altimeter, airspeed indicator, and navigation displays. While performing these scans, they verbally communicate what the instruments give them in information. When changing the altitude from the flight plan, they mutually inform each other by words. One pilot addresses the other to confirm the change as it is being done. The flight deck pilots interact with air traffic control throughout the flight, receiving instructions and providing position reports in the same moment-to-moment sequence. They communicate with controllers over the radio, relaying instructions and clearances to one another. The pilot monitoring may receive an altitude clearance from ATC and then inform the pilot flying, "Cleared to climb to 30,000 feet." The co-pilot, receiving the information immediately, tells the pilot flying and confirms the received information from the air traffic controller. Even though the pilot flying has also heard the communication between the colleague and the tower in his headset, the pilot flying allows the pilot monitoring to give or pass on the information. The pilot flying might not have heard correctly, as he has been occupied with something else. The protocol sequence is that the pilot talking on the radio passes the information from the radio conversation on to the pilot flying. In the event of an emergency, pilots use the same type of protocolled sequencing. They follow established procedures and checklists to address the situation. One pilot may initiate emergency protocols while the other communicates with ATC and assists with troubleshooting or executing emergency manoeuvres.

Throughout the flight, a specific set of actions is undertaken in a particular sequence of tasks and actions to exchange critical information through mutual confirmation to ensure that the various steps have been taken in the correct sequence. When a specific task has been conducted and understood to be conducted and completed, the crew can proceed to the next item on the list.

Pilots use a combination of verbal speech, visual cues, and standardized procedures to orient each other and coordinate their activities. In some phases of the flight, they have different tasks to conduct separately, or in combination with another set of tasks that must also be performed in a particular manner. Then, the pilot doing the task informs the other pilot and proves the completion by using gestures pointing at the instrument, confirming that the task has been done. In this way, the pilots actively complete each sequence together by mutually confirming one another.

Like we saw in the opening chapters, the pilots follow a standardized checklist to ensure that all critical tasks and systems are properly checked before starting the aircraft's engines. Each item on the checklist is called out by one pilot, and the other pilot confirms its completion or status. This dialogue helps in cross-checking and ensures that both pilots are aware of the progress and the status of each task. The sequence described in Chapter 3 is an illustration of how, ideally speaking, pilots - on a moment-to-moment basis - coordinate their individual activities to work together and perform a flight, routinely orienting and achieving this sequential nature in their work.

5.2 TO STAY AHEAD OF THE CURVE

The safe and efficient operation of an aircraft is created through the action, communication, and interaction in the cockpit. The pilots must have the capacity to think and act as one cohesive unit. By shared attention, they stay ahead of the curve, anticipate challenges, and make informed decisions about the flight they are undertaking.

Through the constant creation of a unity of thinking and acting, the pilots share a mental framework and model the flight they are about to carry out. They join together their individual skills into a common understanding of the flight operation. This is possible because pilots do not share opinions, but "a form of life" (Wittgenstein, PI, § 264). When pilots on a moment-to-moment basis share what they hear, see, and do, they share a cognitive representation of what they are doing. By this intense sharing of what they do, the pilots make accurate predictions of the flight and effectively coordinate their joint efforts. The pilots in the cockpit are acting within a synchronized framework, sometimes even without explicit communication on what they do as they act, interpret, and respond to the various situations with which they are both familiar. Their shared framework of the flight and the skills they share allow them to anticipate each other's actions. This fosters a synchronized thought process. Their use of the shorthand language works perfectly well because both are fully competent. They share their familiarity with the practice executed. Wittgenstein's famous builders in PI's § 2 communicate like the pilots in the cockpit, and for the same reason. On the building site, the one-word orders shouted out, such as "block," "pillar," etc., by the master work as orders because the assistant understands what his master does. The assistant understands what his master is saying because he knows what he is doing (Nergård, 2019). There is nothing lacking, and Wittgenstein asks us to see the shorthand language on the building site as a complete language (Wittgenstein, 1953, § 2, § 6, § 241).

The interaction between the pilots in the cockpit consists of a continuous sharing of information about the flight as it proceeds. They are monitoring instrument readings and exchanging information, such as weather updates, air traffic control instructions, etc. All efforts are made in order to reduce the risk and avoid errors. Professional skills and a common readiness and framework make the pilots able to predict the steps of the flight and adapt to each other's perception of it. In the cases where they need to make decisions, they share the background for them.

Throughout the flight, the pilots extensively make use of cross-checking techniques to verify critical information and decisions. By cross-checking each other's acting, they continuously act as "one head." This cross-checking dialogue took place after take-off in a Boeing 737-800:

Pilot A: "Positive rate, gear up."
Pilot B: (Raises the landing gear) "Gear up."
Pilot A: "Flaps up."
Pilot B: (Retracts the flaps) "Flaps up."
Pilot A: "Speed check."
Pilot B: (Verifies the airspeed indicator) "Speed checked."
Pilot A: "Heading select, engage."
Pilot B: (Engages the autopilot and selects the desired heading) "Heading select, engaged."
Pilot A: "Autopilot, on."
Pilot B: (Activates the autopilot) "Autopilot, on."
Pilot A: "Heading captured, altitude select, engage."
Pilot B: (Verifies the captured heading and engages the altitude hold mode) "Heading captured, altitude select, engaged."
Pilot A: "LNAV (Lateral navigation, engage)"

Pilot B: (Engages the lateral navigation mode) "LNAV, engaged."
Pilot A: "After take-off checklist."
Pilot B: (Verifies the completion of each item on the checklist) "After take-off checklist, completed."

(Observation on flight, 2021)

Each section in the dialogue above may vary depending on the airline's procedures, but the overall concept of cross-checking is universal. By actively verifying each other and confirming the desired results, the crew ensures that critical tasks and settings for the phase are correctly completed and monitored. Involving both pilots in the process, errors are minimized.

The sequenced communication keeps the pilots engaged and involved in every aspect of the flight, which is a critical component of safety. When pilots continuously keep one another in the loop, they stay ahead of the aircraft:

> Safety is the ability to manage the workload connected with the single tasks in the flight. At the same time your thoughts should be everywhere else. In front of the aircraft. In your mind you should be ten miles ahead of your fan. I often say to inexperienced first officers that the real safety issue is the ability to be on the top of every situation and be able to identify potential hazardous situations. You need to see the big picture in your flight but also have the big picture on what is going on around you. The ability to think at the same time as you act and are in control is both the same and different challenges. Some elderly pilots use a lot of mental capacity to plot in data in the Flight Director when it was introduced. They didn't have a "natural relation" with computers. They learnt to fly in a time where computers were "alien to the cockpit," another experienced pilot explained.
>
> *(Interview with pilot "Per," 2021)*

Being ahead of the curve turns the pilot's readiness to anticipate and prepare for uninspected challenges. Weather changes, airspace congestion, and mechanical issues can significantly impact a flight's progress. By staying proactive, they monitor weather patterns, review their flight plan, and assess their aircraft performance limitations according to changing weather conditions: "This enables us to detect any deviations or anomalies promptly and respond effectively to unexpected events and maintain control of the aircraft," Captain "Per" explained (Interview, 2021).

In their moment-to-moment dialogues, the pilots distribute their workload, freeing up mental capacity to focus on monitoring the overall situation. This sharing and delegation process enables them to think collectively and stay ahead of the evolving conditions: "Everyday life in the cockpit is not constantly looking for malfunctions and errors but is about creating the calm in which we have the overall picture" (Interview, 2021).

5.3 SITUATIONAL AWARENESS – "FLYING FROM THE GALLEY"

Not all flights are ahead of the curve and may be quite the opposite. "Flying from the galley" refers to the galley area in the aircraft where the preparation of food and beverages to be served to passengers takes place. The galley is often situated just behind the cockpit or at the back of the aircraft. When pilots use the phrase, it means they are behind the situation where the flying takes place - in the cockpit: the crew is not mentally present. On the first flights in a new aircraft type, the inexperienced pilot might feel like the aircraft is flying without the pilot being able to be on top of everything.

When Air France Flight 447 crashed into the mid-Atlantic Ocean in 2009, the tragic accident was caused by pilot failure. The Airbus A330-203 encountered severe weather conditions, including a high-altitude thunderstorm, while flying across the Atlantic. The aircraft's pitot tubes, which measure airspeed, became obstructed by ice crystals and caused unreliable airspeed indications. In response, the flight control computers disengaged the autopilot and auto-thrust systems, which led to a situation where manual flying was required. However, the flight crew did not realise the situation, probably because they were not adequately trained, or they were not prepared to handle the scenario of unreliable airspeed indications and loss of automation. As the situation unfolded, the pilots became disoriented and failed to recognize the impending stall. Trying to master the situation, they made incorrect control inputs, and there was a lack of effective coordination and communication in the cockpit. The aircraft eventually entered a deep stall, where recovery became extremely challenging. Despite efforts to regain control, Air France Flight 447 crashed into the Atlantic Ocean on its way from Brazil to France, resulting in the loss of all 228 passengers and crew members. The accident highlighted the norm in the pilot community of staying ahead and not being behind any situation.

Pilots themselves have only one recipe for the ideal: "If you think you are finished training, and have nothing more to learn, you are an accident waiting to happen" (Pilot interview, 2021). Regularly updating knowledge, participating in recurrent training, and improving proficiency are essential to ensure a shared presence in the situations that might occur during a flight.[3]

The term "Situational awareness" was introduced in the 1980s by Mica Endsley, a cognitive psychologist who was working with the United States Air Force to study the role of human factors in aviation accidents. Endsley's work focused on pilots' capacity to make effective decisions in complex environments. She defined situational awareness as "the perception of the elements in the environment within a volume of time and space, the comprehension of their meaning, and the projection of their status in the near future" (Endsley 2015).

Situational awareness involves not just perceiving information, but the overall understanding of the context and the making of predictions about what might happen next during a flight. Endsley's point was to address the ability of the individual or team to perceive and understand the current and future state of the flight environment, and to anticipate changes and make decisions based on that judgement and anticipation. Being ahead is the pilot's awareness not just of their surroundings but also their own ability to get and interpret adequate information to make informed decisions.

Endsley elaborated her concept by dividing it into three levels: perception, comprehension, and projection. Perception involves the collection of information from the external environment, instruments, and through the senses. Comprehension involves the interpretation of that information to understand the current state of the flight environment. Projection involves the anticipation of future changes in the environment and the ability to make decisions based on that anticipation. Endsley's research emphasized the importance of situational awareness in preventing accidents and improving safety in aviation. Since then, the concept has been widely adopted in other fields, including safety science and emergency response. Today, situational awareness is increasingly considered a crucial component of effective decision-making, research, and training in different industries.

Situational awareness has a significant impact on aviation research and is currently regarded as critical for safe and efficient flight operations. The concept is used to improve skilled communication about the handling of aircraft under shifting circumstances and other relevant information about safety.

The importance of communication in establishing situational awareness was highlighted by several high-profile aviation accidents, such as the Tenerife disaster between two Boeing 747 aircraft in 1977. In this accident, miscommunication between air traffic control and flight crews led to the collision between two aircraft, resulting in the deaths of 583 people. Several factors caused the accident, but the key elements were the breakdown in situational awareness among the flight crews involved.

On the day of the accident, adverse weather conditions played a part. Heavy fog covered the airport and significantly reduced visibility. The pilots' visual assessment of their surroundings was severely limited. Due to a bomb explosion at Gran Canarias Airport, multiple diverted flights were rerouted to Tenerife (Los Rodeos Airport), causing the airfield to become busy. The increased traffic and the need for multiple aircraft to manoeuvre on the limited taxiways further complicated the overall situation.

Radio communication between the aircraft and the air traffic controllers was hampered by various factors, including language barriers, simultaneous transmissions, and misunderstandings. This led to misinformation and confusion about the location and intentions of the aircraft crew involved. A KLM flight, which was attempting to take off, received clearance from the control tower, but the crew misinterpreted it as permission to begin its take-off roll. The crew initiated the take-off without proper confirmation from the tower, assuming the runway was clear, and thus lost situational awareness of the Pan Am aircraft on the same runway. The foggy conditions made it difficult for both flight crews to see each other manoeuvring on the runway or fully understand the imminent collision risk. The absence of visual cues further eroded their situational awareness. These factors combined created the disastrous misinterpretations in both crews and led to the collision between the two aircraft on the runway. The tragedy was the deadliest accident in aviation history.

The Tenerife accident, like many other incidents and accidents, has been re-interpreted and pointed to as a case of absent situational awareness. The aviation community has implemented significant improvements in training, protocols, and technological advancements. These measures aim to prevent accidents by improving the pilot's situational awareness during flight operations.

One of my pilot informants explained:

> When we fly, we are constantly on the lookout for situations that needs to be addressed and problems to be solved. If someone were to observe the acting in the cockpit, they probably see us as a pair of ducks. During a flight we are contagiously working on our situational awareness. We look out for cues from the radio, scanning our instruments, the radar and so on. During the flight we are building a picture of what is going on around us. To the outsider it might appear stupid that to repeat observations and short-hand messages over and over again. Regardless how many times I have been on duty with the

> same pilot in the seat next to me, regardless of how many legs we have done on the same day, or how many times we have landed at the same airport, we follow the procedures and the scheduled briefings. We repeat and do the same each time. The reason is of course that what happened last time do work as a blueprint for the next.
>
> *(Interview, 2023)*

Each landing and take-off is entirely a new situation, even if they are based on long-term experience. The entire reliance on experience is a source of both success and failure. The accuracy of the detailed procedures during a flight is there to prevent the experience from taking over and replacing the awareness that a pilot needs for each flight. The long pilot experience is the foundation of each flight operation. Any flight is dependent on it. Still, it always needs a second opinion (Figure 5.1).

> We do whatever we need according to standard operating procedures. In that sense the procedures are a constant and our building blocks in the image of what is going on around us. When we are flying, when we are airborne, we cannot just simply pull the brakes. We are flying in a rushing speed. That is why the image is constantly changing. We need to always understand what is going on around us. During a flight, there are countless inputs associations as things you need to cope with and portray. As your experience grow, you learn to eject some of the inputs that is irrelevant. For example, when you are flying in a tight airspace the input from the radio is for a newcomer quite overwhelming. But as you learn, you eject all the call signs that is not yours. When I'm on the radio, it's like I don't pay attention to any other call sign then those who start with "Scandinavian." Then the rest of it is all about the flight number.
>
> *(Interview, 2023)*

FIGURE 5.1 Creating safety.

(Photo: Christian Aasen Stratton.)

5.4 SAFETY AS A DIALOGUE AMONG PROFESSIONALS

Accident investigators have increasingly become aware that a critical factor in several incidents and accidents is pilots' loss of sense of what is going on around their performance of the aircraft. The state of lost control of a critical moment in a flight has, in several cases, been identified as the direct or indirect cause of disasters.

Situational awareness is the utmost perception needed in the cockpit. When pilots are flying in a congested airspace, they need to combine their individual situational awareness of the position and altitude, the location of other nearby aircraft, and any potential conflicts or hazards. The shared awareness allows the crew to maintain safe and timely adjustments. The routines involve them effective with another, and any misjudge may be corrected.

The shared situational awareness is of the utmost importance in the cockpit. Pilots complement each other, while their individual situational awareness at the same time is needed to understand the immediate surroundings and make decisions. The shared situational awareness comes from both sources, as they enhance it in communication to create as a shared understanding of the overall situation.

Pilots in a multi-crew setting must develop and demonstrate their situational awareness by training in context on moment-to-moment routine interaction. The importance and value of working in teams have increased in pilot training over the last decades. The "one-man-band-pilot"[4] does not exist any longer. In line with technological development, modern cockpits have become an ever more information-rich environment. The new generations of aircraft are highly computerized and automated. Flying in modern aircraft means managing the technology of aircraft systems that entirely belong to a team.

If a captain is flying with an inexperienced first officer, and the first officer is the pilot flying, the captain will listen actively to the first officer's descent and pre-landing briefing. Through the briefing, the captain will actively make up his/her mind about how much support the first officer will need.

In the planning phase and the briefings, the pilots actively construct a mental model[5] of the flight they are about to do (Dekker, 2014) Meeting up in a briefing setting before the flight, they go through the weather information, technical information about the aircraft, and knowledge of the destination. Everything that can be of interest for the safety of the flight is revised. In leading airlines, most of the information needed for the planning of the flight is provided by others than the pilots themselves: the load sheets, pax info, route planning, fuel estimates, frequencies, etc., all information provided to the pilots by professional planners. For most pilots in large airlines, the pre-flight brief is merely a question of going through all available information and implementing it when they sit down and use the information provided for them as they plan the flight leg(s) from start to finish.

It is in this planning process that pilots make their essential decisions. When the crew plans their flight, they also plan what procedures to use and follow in the flight they are about to execute. This planning process is the practical aspect of creating a mental model (Orasanu, 1994) of the flight. Both pilots must take an active part in the planning because the planning is the active construction of the mental model to be applied. Both pilots must share and perceive this mental model as "owned" (Dekker, 2014).

The research approach has been to treat pilots' mental models and their situational awareness as an individual and perceptual phenomenon based on the information available to pilots. More attention must be given to the social aspects of situational awareness. The industry and researchers now recognize that improving pilots' shared situational awareness will mean more than providing better displays and brighter and louder alerts. There is considerable research and practical work to be done to better understand situational awareness as a team issue.

Even if pilots' communication is a critical role in the flying performance, communication skills have yet to be examined more closely and in the aviation context. Communication is frequently mentioned in the aviation literature, but how pilots can improve their communication skills and what may happen if they do not need to be clarified and better understood in a business with accelerating complexity on all levels.

Establishing a good working relationship with the pilot community is the natural first step in creating flight safety. There needs to be substantial knowledge bridging the gap between pilots' first-hand experience of safety matters and the dependence of passengers and the industry on it.

5.5 AWARENESS OF COLLECTIVE MINDS

Captain "Per" displayed in the previous paragraph the importance of the verbal interaction in the cockpit. This is a key to flight safety and is carried out in a very practical manner. The pilots use just as much body language, gestures, facial expressions, head movement, and tone of their voices as we can observe in other advanced professional practices, where doing is as important as talking. Pilots are talking in their acting, just like Wittgenstein's builders. This interactive acting requires a skilled understanding of the work, just as the communication in the cockpit does, with one implying the other.

The standard operating procedures, such as the checklists, are merely the words that are used to describe their actions in a different sequence of the flight. The meaning of them is what they create in the interaction, making the flight work and what they add to the joint adventure.

The interaction within a flight crew is the overall challenge, and it can be carried out just by the capacity of a performing team. The relevant literature contains several studies underlining the importance of interaction between the pilots. Flying an aircraft requires high levels of cognitive skills, practical performance, effective decision-making, and efficient teamwork.

Pilots must actively share their individual expertise and experiences with each other. Each crew member contributes to the collaboration by sharing skills and observations. Crew member brings a unique set of skills, knowledge, and perspectives to the team. By exchanging ideas on a largely fixed challenge, the crew can tap into the collective intelligence of the group. Interactions within a flight crew contribute to the management of the task and the workload. By coordinating and dividing responsibilities, crew members optimize their individual cognitive loads, ensuring that no one becomes overwhelmed. This sharing and distribution of the tasks appropriately allows crew members to allocate their resources efficiently and remain focused on critical aspects of the flight.

Several studies on decision-making have underlined the importance of a collaborative problem-solving culture that fosters creativity, critical thinking, and adaptability, enhancing the capacity of the crew as a whole.

By actively engaging with one another, team members facilitate each other in the team to speak their mind and pool their collective efforts towards mitigating risks. The seamless coordination and synergy derived from team interaction are the main sources of a safety culture that promotes proactive decision-making and error prevention.

NOTES

1 Several researchers have suggested that crew members have shared mental models of how to fly airplanes. The better the model, the better the crew performs. Cannon-Bowers and Orasanu refer to the shared understanding as "Shared mental Model," which states that the basis of good performance, good cooperation, and interaction is that communication must create a common situational awareness within the flight crew

> Having a shared mental model does not mean that all crew members think exactly the same, but it does increase the likelihood that the pilots on a flight deck are moving in the same direction. All have the same understanding of what the problem is. All know what is to be accomplished and both know how it is to be accomplished.
>
> *(Cannon-Bowers & Orasanu: Flight Psychology, 1990)*

2 In most airlines, SOPs (Standard Operating Procedures) are designed so that pilots routinely engage one another in their activities during any stage of a flight. Ahead of the pre-flight sequence, the pilots gather to discuss the flight plan, weather conditions, and any specific considerations. They share information about the aircraft's systems, fuel levels, and cargo to ensure that both pilots share a common view of the state of the aircraft, the route, and the flight ahead. In the cockpit, they conduct a series of checks to verify that the aircraft is in proper working condition.

3 According to EASA regulations (Part-FCL and Part-OPS), commercial airline pilots are required to complete recurrent training and checking at least twice per year, typically in a full-flight simulator. These Licence and Operator Proficiency Checks (LPC/OPC) ensure compliance with technical and procedural standards. In addition, most operators provide supplementary courses in Crew Resource Management (CRM), safety, and emergency procedures, making recurrent training a regular part of professional life.

4 The "one-man-band pilot" historically represents an era in aviation when pilots operated independently, often resistant to collaborative practices or shared situational awareness. These pilots were known for their self-reliance, managing all aspects of flight without the input or assistance of others. This mindset was a product of the time, shaped by limited technology, simpler aircraft, and a culture that celebrated individual skill and autonomy. However, as aviation evolved, the complexities of modern flight demanded cooperative crew dynamics and shared situational awareness to ensure safety and efficiency. The one-man-band pilot, characterized by their reluctance to embrace teamwork, became obsolete as multi-crew operations and standardized procedures became the norm.

5 In Endsley's concept of situational awareness (SA), the "mental model" refers to the internal representation or framework that an individual uses to understand and predict the functioning of a system or environment.

6 Communication and Flying Skills

Communication derives from the Latin verb *communicare* and means "to share" or "to impart." It is the state of being in a common language, culture, and values. Communication belongs to all human activity and social interaction. It is essential to human survival and development.

6.1 ALL IS COMMUNICATION

Communication is the lifeblood of human interaction. Also, in aviation. It forms a specific system, fitting together scientific knowledge and everyday life matters in flight operations. Communicating can be intended or unintended; nonetheless, humans continuously communicate in all human settings. Every human act contains a message. Humans communicate not just when speaking or intentionally conveying a message but also by body language, facial expressions, tone of voice, and most other nonverbal cues without being accompanied by spoken words. Silence also communicates. Not speaking may convey messages, disapproval, or disinterest. Silence may even be a deliberate strategy for communication in human practice if which is a part of (Meløe, 1974, 1988, Merleau-Ponty, 1994)

Communication is not solely about the individual but also belongs to shared expressions. In aviation, communication is bound to a wide range of formal and fixed forms about technologies and protocols, all designed to facilitate the flow of information exchanged, shared, and transmitted between the actors involved in flight operations. They enable pilots, air traffic controllers, ground personnel, and other personnel to coordinate critical information. In the web of communication, everyone has a role to play, not unlike a theatrical play. The expectations of each individual actor and the consequences of failures can be far more fatal than in any role played on a stage (Goffman 1974).

Since the pioneering period of commercial aviation, advanced communication systems have been developed. The evolution of advanced communication systems has been deeply influenced by the aftermath of tragic accidents, many of which were rooted in fatal human errors. A striking early example is the 1972 Eastern Airlines Flight 401 disaster. While focused on troubleshooting a malfunctioning landing gear indicator, the flight crew failed to maintain a comprehensive understanding of the aircraft's flight state. Unnoticed, the autopilot disengaged, and the aircraft began an unmonitored descent into the Florida Everglades. The crew's communication broke down, and the first officer and flight engineer did not alert the captain to the altitude loss, ultimately leading to a crash that claimed 101 lives. The investigation pointed to poor communication, a narrow fixation on a single issue, and the crew's inability to maintain overall vigilance as critical factors in the tragedy.

DOI: 10.1201/9781003725862-6

The Eastern Airlines Flight 401 disaster spurred significant awareness of communication challenges within the aviation industry. From the late 1970s, a pioneering group of researchers at NASA initiated a systematic approach to improving aviation communication, particularly the onboard communication practices between pilots. Had the flight crew members shared pertinent information, cross-checked their instruments, and engaged in clear and open dialogue, they might have identified the impending disaster and taken appropriate action to prevent it.

A similar breakdown occurred six years later during United Airlines Flight 173's approach to Portland, Oregon. While investigating a landing gear malfunction, the crew became so absorbed in diagnosing the issue that they neglected to monitor fuel levels. Additionally, they failed to clearly communicate their deteriorating situation to air traffic control (ATC) or request emergency assistance. By the time they realized the severity of the fuel depletion, the engines had flamed out, forcing a crash-landing in a suburban area. The crash resulted in 10 fatalities and injuries to 175 others. Investigators found that the crew's lack of coordination, prolonged focus on a single problem, and the captain's dismissal of the first officer's concerns were key contributors to the accident.

The investigations into these incidents highlighted failures in maintaining a broad and accurate understanding of the operational environment - an aspect now recognized within the framework of situational awareness. These tragedies underscored the necessity of structured communication, effective teamwork, and shared vigilance among flight crews. These principles later became the foundation of Crew Resource Management (CRM), a systematic approach formalized by NASA research in the late 1970s. CRM introduced tools and strategies to foster collaboration, encourage open dialogue, and ensure flight crews collectively monitor and manage the aircraft's state, preventing the recurrence of such failures.

6.2 AVIATION CODES AS A LANGUAGE

A flight communication system organizes, regulates, and realizes human actions. It developed in the context of aviation engineering and technology, and became a part of a specialized communication system, like other professional human activities have developed their language practices in other branches.

Aviation language has a specialized jargon and terminology, abbreviations, and standardized phraseology. This language includes specific phrases, acronyms, and standardized protocols used by pilots, cabin attendants, air traffic controllers, and ground personnel. It also includes standard radio phraseology, instrument approach procedures, and traffic separation rules to ensure that messages are transmitted accurately and efficiently. It also includes visual communication systems, procedures, and digital communication technologies.

This language and the communication in which it is issued is a code system with the purpose of making orders and commands fully accurate and complete, as any other language in any other practice with the same logical order (Wittgenstein 1953). It is used within a working community where the involved are fully acquainted with what is going on and why. The verbal utterance of a certain sign is the normal use of this strictly coded language and communication among crew members during a flight. The coded messages have a complete meaning when they are used. We may say that the practice they are part of is their dictionary. This language is developed, made for, and spoken just there. As specific protocols and procedures are followed during flight operations, this phraseology is also used in radio calls, position reports, clearances, and acknowledgments.

Pilots and most other involved, communicate within this unique linguistic landscape marked by a distinct set of language and communication: "As pilots, we have a specialized language, a set of terms and phrases. The term 'go-around', for instance, holds a specific meaning. It means a procedure to abort a landing attempt." Pilot Henning underscored that "go-around" is not merely a word; rather, it encapsulates both a phrase and a well-defined procedure within the aviation context - a procedure employed to abort a landing attempt (Interview with "Henning," 2019).

In the high-stakes environment of piloting, split-second decisions may have far-reaching consequences. The clarity of short terms is evident. The context in which they are articulated, whether it be in routine or emergency situations, determines a precise meaning.

The shared understanding of specialized terms forms the bedrock of communication. The meaning is not derived from the words themselves but from the shared experiences accompanying their use. Pilots' ability to communicate with precision and clarity hinges on this linguistic framework. The term "go-around" is imbued with a depth of meaning that extends beyond its literal definition. This language system contains a set of vocabulary, where the emphasis is on operational efficiency, rather than linguistic itself. The dialogues between the stakeholders have its emphasis on purposefulness and a fast pace. The longer a communication sequence lasts, the less time is left for other necessities.

"Adequate acting means good understanding," is Meløe's reflexion when addressing any practical deed or performance (Meløe 1979). It certainly holds for the language and acting on a flight deck. We may say it is about the economy of language and communication.

The cockpit language makes communication accurate and precise. Outside, it may appear confusing. Becoming an insider demands this language. To be familiar with it requires being an insider to the practice of flying gained from a globally recognized education system for pilot training. Through this training in advanced pilot skills, the pilots share with other professionals in medicine, science, and other practices, combining advanced knowledge and practical skills. The professional skills cannot be either theoretical or practical; they must be both and fit together.

6.3 LANGUAGE AS A SAFETY ISSUE

English is the universal language for aviation communication. The International Civil Aviation Organization (ICAO),[1] an agency of the United Nations, established English as the standard language for international aviation communication. It promotes uniformity and consistency in aviation operations worldwide.

As English was established as a lingua franca, all pilots and air traffic controllers with different linguistic backgrounds communicate in a language that is not their mother tongue.

This has not come without setbacks and sometimes also disastrous results. In the globalized aviation, pilots and air traffic controllers with diverse linguistic backgrounds and not all speak Queen's English. Pilots and air traffic controllers are native speakers of one language and speak Aviation English while on duty. Using aviation language normally also means good English, but there are exceptions.

Not surprisingly, language has been identified as a causal factor in safety-related incidents and accidents. A major example, discussed in Chapter 5, is the KLM-Pan Am Tenerife Airport Disaster (1977), where communication barriers contributed to the collision of the two Boeing 747s on the runway. The Tenerife disaster was primarily caused by miscommunication between the flight crews of the two aircraft and ATC. Dense fog and limited visibility prevented the crews from seeing each other on the runway, compounding the situation. The KLM captain mistakenly believed he had been cleared for take-off after acknowledging a departure clearance, a critical misunderstanding exacerbated by non-standard and ambiguous phrasing. For instance, the KLM first officer's statement, "We are now at take-off," was misinterpreted as an active take-off roll. ATC's response, "Stand by for take-off," failed to clarify this misconception decisively.

The Tenerife disaster emphasized the need for standardized phraseology, proficiency in English, and improved communication protocols to mitigate the impact of language barriers (in aviation). As a result, advanced English language proficiency is a requirement for pilots and air traffic controllers in most countries and companies, but not in all.

The aviation communication system is far from straightforward. The vast cultural and contextual diversity in recruitment is in itself a contribution to "human errors" as is well-documented in various reports and studies. Human errors, often labelled as "the human causal factor," are a major part in numerous air accidents (Dismukes, 2009). Still, the communication in aviation is not only to elucidate best practices but also requires extensive examinations of miscommunication.

There is also a growing awareness of cultural factors contributing to aviation accidents (Cookson, 2019). Remarkably, human factors are estimated to be responsible for a staggering 70–80% of such incidents. Several comprehensive studies have made a strikingly similar revelation: approximately 70% of all aviation accidents can be attributed to crew coordination and communication issues (Chatzi, Martin Bates & Murray 2019). Even more striking is the fact that, for years, the most common type of air disaster has been the so-called CFIT accidents.[2] These incidents involve an unintentional collision with terrain while the aircraft remains under the positive control of the flight crew. The aircraft itself typically has no mechanical issues. Rather, the tragedy stems from the pilots' inability to maintain situational understanding, leaving them unaware of the impending disaster until it is too late.

Errors, mishaps, lapses, and fatal accidents have been the most common factors. It is not the communication itself that often causes the accident, but disruption in the communication process (Parret 1994). Miscommunications may occur due to failure in the transmission or reception of a message. The "mis-" prefix in "miscommunication" indicates a negation, error, or lack of accuracy in communication. The communication process did not happen as intended, and there was a misunderstanding, mistake, or misinterpretation of the message. The intended meaning of a message given or received does not align with the original intention.

Interpretation is like communication, something we continuously do. The pilots are involved in a continuous process of sensemaking and responding to cues.

They create and construct their understanding of reality in a continuous, ongoing process during the flight. This involves adjustments in response to the dynamic flow of events. The pace of sensemaking aligns with the promptly updated interpretations of any situation on the flight deck (Weick, 1995).

Miscommunication differs from misunderstandings, often manifesting as alterations or distortions of an original message. These changes may occur inadvertently, especially when crucial details are omitted during the transmission process. Misinterpretations can result from wrong assumptions about mutual understanding, hindering adequate responses due to the misinterpretation of instructions or failure to convey crucial information. Challenges may escalate, such as when English proficiency issues contribute to interpretation turning into a potential hazard. Aviation history is replete with examples of miscommunication or misunderstandings, especially in critical moments.

The Swiss Cheese Model, introduced by James Reason, is widely used in aviation education as a risk management tool. It illustrates how accidents occur when multiple layers of defence develop vulnerabilities, represented as holes, through which a hazard can penetrate. Each layer symbolizes a safety barrier or control, highlighting that no single layer is infallible. Accidents happen when the holes in these layers align temporarily, allowing a hazard to progress through the system unimpeded. Culture and language serve as essential defence layers that typically help prevent accidents. However, cultural nuances and language intricacies can sometimes introduce complexities or misunderstandings that contribute to the misalignment of communication, creating additional vulnerabilities within the system.

The challenge of communicating in a foreign language is exemplified by the tragic events of Avianca Flight 52 on January 25 in 1990. Flight 52, carrying 149 passengers from Bogotá, Colombia, to New York, encountered adverse weather conditions and air traffic congestion, leading to a forced holding pattern off the coast near New York. As the Boeing 707 circled for nearly an hour and a half, fuel levels dwindled, setting the stage for a critical situation upon arrival at Kennedy Airport.

Upon reaching the airport, the challenges intensified. Due to fog and wind, only one runway was operational for the 33 planes attempting to land every hour. The already delayed Flight 52 faced further setbacks as the aircraft ahead failed to touch down. With fuel reserves running critically low, the situation became desperate.

After the missed initial attempt to land, Flight 52 was assigned a landing pattern that it had insufficient fuel to execute. While the crew struggled to manoeuvre the aircraft, its engines flamed out successively. The Boeing 707 tragically crashed into the village of Cove Neck, Long Island, resulting in the loss of 65 passengers and eight out of nine crew members.

The subsequent investigations revealed a failure on multiple fronts. Two miscommunications played a central role in the unfolding tragedy. Firstly, when the aircraft transitioned from regional to local air traffic controllers, the local controllers were not adequately informed about the critical fuel situation and the fact that Flight 52 had insufficient fuel to reach its alternative airport. This lack of crucial information deprived the local controllers of a comprehensive understanding of the dire circumstances.

Compounding the issue, the aircraft's crew failed to explicitly declare an emergency to the local controllers. Such a declaration would have signalled that the plane was in imminent danger of crashing. The absence of this critical communication further obscured the severity of the situation, and the crew's failure to use standardized terminology further hampered effective communication. Despite the urgency of the circumstances, the crew did not explicitly declare a "fuel emergency," leading to a tragic misunderstanding between the pilot and co-pilot.

The National Transportation Safety Board (NTSB), the U.S. agency responsible for investigating aviation accidents, identified the probable cause of the crash as the "crew's failure to manage the airplane's fuel load adequately and their failure to communicate an emergency fuel situation to air traffic control before fuel exhaustion occurred." The investigation also pointed to issues in how the Federal Aviation Administration (FAA), responsible for air traffic control, managed traffic flow and the lack of standardized terminology for minimum and emergency fuel states.

However, investigators from Colombia, representing their national aviation authority, disagreed with parts of the NTSB's findings. In their review of the NTSB's draft report, the Colombian team suggested assigning some responsibility to the air traffic controllers, citing their "inadequate handling" of the Avianca flight.

The testimony by Captain Saul Pertuz in the (NTSB) panel hearings added another point to the discussion about the miscommunication: On the flight deck, the pilots used two languages, English in communication with controllers and Spanish in conversations between the captain and the first officer. The captain questioned how the NTSB could claim that the crew had failed to declare an emergency explicitly. Despite the first officer stating, "We're running out of fuel," the explicit declaration of an emergency did not occur immediately.

"I was very surprised that 'priority' and 'running out of fuel' meant nothing to them," Captain Saul Pertuz testified to the NTSB panel.

Referring to an Avianca route manual used in the training of the airline's pilots, the captain said there is nothing in it "that says we must use specific words."

"In my country and in this country, there have been a great number of (emergency situations) in which the pilot would not use the word 'emergen'" Captain Pertuz pointed out: "He would just tell them what happened."

The first officers' copy of the route manual, recovered from the wreckage of Flight 52, showed that Avianca's pilots had been trained to "declare an emergency . . . and report fuel remaining . . . if the remaining usable fuel supply suggests the need for traffic priority to ensure a safer landing."

Recordings recovered from the cockpit show that the failure to declare an emergency may have stemmed from a misunderstanding and misinterpretations between the pilots and between the pilots and the air traffic controllers: After an initial attempt to land at Kennedy was aborted when the plane sank below the safe glide path in heavy, gusty winds, the first officer radioed controllers, "We're running out of fuel . . . ."

The captain instantly told his first officer: "Tell him (the ATC) we're in emergency."

Sixteen seconds later, the captain repeated his order: "Advise him we are in an emergency."

Four seconds after that, the captain asked the first officer, "Did you tell him?"

Although he had made no mention of an "emergency," the first officer replied, "Yes, sir, I already advised him."

About 30 seconds later, the first officer again radioed controllers that "we're running out of fuel."

"Did you already advise him that we don't have fuel?" the captain asked the first officer.

"Yes, sir," the first officer replied.[3]

Then, in one of the responses that remains a mystery, the first officer told controllers who asked him if his fuel was adequate for the new approach route they had recommended:

"I guess so. Thank you very much." About 7 minutes later, its engines starved of fuel, the plane crashed[4] (Transcripts of the CVR, 2025).

The tragedy surrounding Avianca Flight 52 illustrates the intricate web of miscommunication, misinterpretation, and misunderstanding. This catastrophic event is an illustration of the multifaceted nature of communication challenges pilots may have, encompassing issues such as language barriers, procedural lapses, and difficulties in communication between various stakeholders involved in flight operations.

6.4 THE PROOF OF THE PUDDING

During the last decades, several communication principles and techniques have been implemented in order to increase flight safety. "The real challenge to flight safety is what happens between us," one pilot told. "It has been proven that most incidents and accidents are related to us on the flight deck, and in particular, the communication between us," he continued. "As a captain, I have a huge responsibility to make sure that the communication on the flight deck works" (Interview with Steinar, 2015).

The exploration of pilot roles in the cockpit can be partly portrayed by Goffman's (1974) concept of symbolic interaction, gained through an experiment conducted by Pan American World Airways in collaboration with NASA during the 1970s. Utilizing a full-motion simulator, the study, designed by Dr. Hugh P.R. Smith, a British physician and pilot, underscored the significance of workload management and internal communication over individual piloting skills. The study challenged longstanding aviation traditions, asserting that effective teamwork and communication were vitally important in reducing accidents, nourishing a collaborative cockpit culture.[5]

The dynamics of communication and hierarchy between the captain and first officer have evolved significantly over time. Historically, the cockpit operated under a rigid hierarchical structure, with the captain holding unquestioned authority and the first officer expected to follow directives without challenge. This steep authority gradient often hindered open communication and collaborative decision-making.

Over time, it became the global standard in aviation, influencing training methods and emphasizing teamwork over individual skills. The cockpit gradient,[6] representing power dynamics between captain and first officer (Helmreich & Foushee, 1993), shifted from a traditionally steep hierarchy to a more balanced approach, recognizing the need for effective communication and shared decision-making in ensuring overall flight safety (Alkov et al., 1992; Weick, 1995; Endsley, 2015).

In the 1970s, a series of aviation accidents highlighted the dangers of rigid hierarchies. Investigations revealed that first officers were reluctant to question captains' decisions, even in critical situations, due to the prevailing authoritative culture. For those inside the cockpit, this often meant noticing an error yet holding back, aware that challenging the captain could be seen as insubordination. These findings led to the development of CRM, emphasizing teamwork, communication, and the importance of all crew members voicing concerns. The shift towards collaboration initially met resistance from senior pilots, some of whom dismissed it as "psychobabble."

In the contemporary era, the cockpit gradient embraces a more equitable distribution of power, emphasizing open communication and active participation from all members of the flight crew. This departure from traditional hierarchies acknowledges the diverse expertise within the team, allowing co-pilots to engage dynamically in discussions, express opinions, and contribute to the decision-making process during flights.

Cultural factors also play a significant role in cockpit interactions. In Western airlines, particularly in the United States and Europe, there has been a shift towards flatter authority gradients, promoting a more collaborative environment where first officers are encouraged to actively participate in decision-making. Still, the acceptance of this flatter cockpit gradient faced challenges in certain Asian countries, where cultural norms of hierarchy and respect for elders posed obstacles. In some Eastern airlines, cultural norms that emphasize respect for authority and hierarchical structures can result in steeper authority gradients, potentially impacting communication dynamics (Merrit 1993b, Merrit & Helmreich 1996, Braithwaite 2001a). For instance, in certain Asian cultures, the high-power distance between leaders and subordinates may lead to an authoritarian decision-making style, where subordinates are less likely to challenge or question superiors. Incidents like the 1997 crash of a Korean Air Boeing 747 highlighted the critical role of communication and teamwork in preventing catastrophic outcomes, reinforcing the importance of a balanced power dynamic in the cockpit.

Captaincy in modern aviation reflects a significant evolution from the traditional, authoritarian role of the past to a leadership model defined by collaboration and shared decision-making. The captain remains the ultimate authority on board, yet their role today is anchored in fostering teamwork and ensuring that all crew members feel empowered to contribute to the operational process. This approach aligns with the principles of CRM, which emphasizes the importance of collective input and mutual respect in achieving safety and efficiency.

A contemporary captain balances authority with approachability, ensuring that decisions are informed by the diverse expertise of the flight crew. Rather than issuing directives in isolation, modern captains engage with first officers, cabin crew, and ATC to gather information and consider alternative perspectives. This shared decision-making reduces the risks associated with errors, particularly in high-pressure scenarios, where collaborative thinking can prevent small oversights from escalating into critical issues.

Leadership in the modern cockpit extends beyond technical skill or procedural knowledge. It demands emotional intelligence, including the ability to manage stress, maintain composure, and provide clear guidance under adverse conditions. A captain is also a steward of the cockpit culture, creating an environment where communication flows freely, uninhibited by rank or hierarchy. The cultural nuances present in today's global flight crews require captains to adapt their leadership to bridge differences, recognizing how varied communication styles and hierarchical norms influence team dynamics.

During my fieldwork, I observed a particularly striking example of this modern leadership approach. In an interview with a seasoned captain of an international airline, he recounted a situation when a first officer pointed out a potential issue with an aircraft's navigation system shortly after departure. While the problem was minor, the captain immediately acknowledged the observation, involving the first officer in resolving the situation. The captain later explained, "I always take input seriously, whether it's from my first officer or even the cabin crew. Small things can save big situations" (Interview with Per 2018). Like Captain Ove in the first chapter, Captain Per is a part of the new generation of Captains, who understand the importance of fostering an environment where everyone in the crew and every voice is heard and respected, regardless of rank or experience. For these Captains, team cohesion, mutual trust, and ensuring that even the most minor concerns are given due consideration, are an important part of creating safety on a flight.

In this perspective, modern captaincy embodies a dual responsibility: upholding the authority needed for operational safety while cultivating a workplace culture built on inclusion, trust, and shared accountability. In everyday practice, this blend of authority and collaborative leadership defines the essence of captaincy in leading their crew to meet the highest standards of safety and performance.

6.5 LANGUAGE SYSTEMS CREATE SAFETY

"What the aviation industry has learned from accident investigations, it is basically two things, or should I say, two mechanisms or learning points: 'Closed loop communication' and the concept of a 'sterile' cockpit," one of these modern Captains I interviewed claimed during a flight (Interview 2019). "I say, you are really showing off your intellectual skills," the first officer commented. They both laugh. "No, but you are right, I agree," the first officer replied.

The humoristic atmosphere the first officer introduced, the captain continued: "Don't tell my wife, but over the years we have been taught how to communicate in a particular way, designed to ensure that messages are not just conveyed but also thoroughly comprehended." Through the laughter, he continued, "This is not a mere formality, it is actually a way of communicating we do. It probably does not work in life outside aviation" (when talking with his wife).

Here, Wittgenstein's notion of a "language game" becomes relevant. In the cockpit, words do not merely describe; they perform. Utterances are bound to action and responsibility. Saying "gear down" or "checked" is not the same as casual speech at home. It is a move in a game where safety and lives are at stake. This makes cockpit communication not only a technical practice but also an ethical one, where language is disciplined so that misunderstanding is minimized and accountability is shared.

Closed-loop communication is a structured method of information exchange aimed at minimizing misunderstandings, particularly in high-stakes environments like aviation (Kanki & Palmer, 1993). This process involves three essential steps: the sender conveys a message or instruction, the receiver repeats the message back to confirm understanding, and the sender verifies that the repeated message is accurate. This method ensures clear and precise communication among crew members and air traffic control. By verifying mutual comprehension at every step, closed-loop communication enhances operational safety and reduces the likelihood of errors, making it a cornerstone of what is often labelled as "effective communication". The communication method functions as a safeguard against misunderstandings and helps to ensure accurate message transmission. For instance, during a communication sequence, the transferring pilot provides information to the receiving pilot, who acknowledges the information and may ask clarifying questions to confirm it. This feedback loop allows for the identification and resolution of any potential misunderstandings, reducing the risk of errors.

During the interview, the air traffic controller's voice crackled through the headset, issuing instructions regarding altitude changes and route modifications. The first officer listens attentively, absorbing every word. The first officer confirms his understanding by repeating the instructions back to the controller. Meanwhile, the captain also has his headphones on as the first officer acknowledged the controller's guidance. What followed was a communicational dance of confirmation and clarification. After this sequence, the first officer repeated the instructions to the captain. Though the captain heard the communication sequence, he asked the first officer questions to ensure they had grasped every detail.

"What you just witnessed is a practical example of closed loop communication," the captain commented, receiving a confirming nod from the first officer, who continued:

> As the controller issues instructions regarding altitude changes and route adjustments, I reciprocated by echoing the instructions. It is kind of a ritual, but it is formed as a loop of information. The short chat created a closed circuit of information. This isn't mere repetition - it is a way of securing it. Had we the same understanding of the instruction?

He continued:

> Take a hypothetical example: the skies over Heathrow Airport are buzzing with activity. A flight crew, while communicating with air traffic control, receives an instruction that could mean the difference between a safe landing and disaster. The pilot's response, an immediate echo of the command, ensures that the crucial information is not lost in transmission. You repeat this instruction back to the controller, ensuring that you've heard correctly. But that's not where it ends. You extend the information to your co-pilot, who also verifies your understanding and may inquire about specific details.

This loop of information sharing minimizes the risk of miscommunication and error.

When the cockpit is "sterile," all communication is solely focused on decision-making and the safe operation of the aircraft: "non-essential conversations or activities that might divert your attention are off-limits," the first officer underlined. The sterile cockpit rule aims to minimize distractions and enhance flight crew situational awareness during critical stages of the flight.

> When we are a "good crew" like today, and we enjoy each other`s company, it is easy to let the atmosphere in the cockpit fool us, and we let our guard down. In the enjoyment of good companionship, it is sometimes hard to stop chatting away about this and that. And, by all means, creating a good athmosphere matters. But the sterile cockpit rule is the reminder that at certain stages, the talking must stop and the attention narrow. It is not a rule in the strict sense, like silence beginning at a precise altitude during descent. It is more of a gold standard, the captain explained.

The backdrop of the atmosphere discussion of the sterile cockpit is all but joyful. An example of an air disaster where the violation of the sterile cockpit practice and norm was a factor is the crash of Comair Flight 5191 in 2006. The flight, operated by a Bombardier CRJ-100 regional jet, was preparing to take off from Lexington in Kentucky. During the taxi and take-off roll, the flight crew engaged in a conversation, distracting their attention from confirming the correct runway for departure. The pilots failed to notice they were on a much shorter runway intended for the aircraft that overran the runway, crashed, and caught fire. The accident claimed the lives of 49 of the 50 people on board. The NTSB[7] investigation identified the violation of the sterile cockpit rule and the crew's failure to maintain proper situational awareness as contributing factors to the accident. The distraction caused by the cockpit conversation compromised the crew's ability to cross-check critical information for the selection of the correct runway.

The sterile cockpit rule draws a sharp operational line: for a period of time, communicatioin is reduced to what keeps the aircraft safe. It is a discipline of silence, but also a discipline and attention. Yet the moment we move out of that phase, the cockpit opens again. The same two people who just sat in a tightly focused operational bubble must shift back into ordinary human interaction – and this transition is never trivial.

Outside the sterile window, communication returns as the central tool of cooperation. The cockpit becomes what it always is in practice: a small social world where two indivuduals navigate both a technical system and one another. Atmosphere, mood, tone of voice, personal habits, cultural expectations. All of it becomes part of the operational tast the moment the sterile cockpit ends.

The flight deck is the primary site for pilot-pilot communication. The cockpit is not a social vacuum. Within the limited space, pilots share the flight deck according to their roles, the captain on the left-hand side and the first officer on the right. Through intensive studies of ergonomics, the flight deck on commercial aircraft is designed in a way that both pilots can reach all the systems and the instrument panels. Even if the captain and first officer have some of the same instruments on their side, they share almost all the cockpit instruments. The systems on each side of the cockpit are within reach for both. The pilots sit very close in one another's personal and intimate space (Hall, 1963). When taking off in a Boeing, the pilots' hands often touch, and if the captain is to select "Gear up," the captain is well within the first officer's "private zone."

The physical distance between the pilots influences the communication between them: "In the cockpit we sit so close that it is almost as if you can smell the mood of the other. You can feel if the other is tense, nervous, happy or sad without being said" (Interview, 2016).

Unwillingly, the cockpit becomes a dual space - a meeting point for diverse cultural norms and personal-space expectations. Pilots operate inside one another's personal and social zones, often closer than they would ever be in everyday life. The challenge is to handle these zones in a way that supports, rather than disrupts, the direct social interaction that lies at the heart of cockpit work. It is through this dialogue – this mind-to-mind communication – that pilots create the knowledge needed to solve the challenges of the flight.

Sometimes this is straightforward. Sometimes it is not. The creation of a solution can itself become difficult, either because communication falters or because the situation has no clear procedures to follow. When a dialogue becomes productive, it can open a space for what Paul calls self-distanciation - the ability to step back from one's habitual ways of acting (Paul 2011). Such moments can be transformative, depending on how deeply the pilots engage relationally with one another. When this happens, they are more likely to take responsibility not only for the joint task they face but also for the relationship that makes coordinated action possible.

For example, during a flight, a captain and first officer may encounter an unexpected weather phenomenon that is not directly addressed in the flight manual. In such a situation, a productive dialogue might begin with the captain acknowledging the limitations of standard procedures and inviting input from the first officer. Through this exchange, both individuals distance themselves from the assumption that the captain alone must provide all the answers. Instead, they collaboratively explore unconventional solutions. This distancing from traditional hierarchical roles enables them to combine their perspectives and experiences, leading to conceptual change through processes such as conceptual expansion (envisioning an alternative route), conceptual reframing (reassessing the weather as an opportunity to test a new approach), and conceptual combination (integrating their ideas into a shared plan).

When these new distinctions are intersubjectively accepted - when both individuals agree on the proposed solution. They create new knowledge that resolves the immediate problem and, potentially, informs future situations. Such instances exemplify how self-distanciation within a productive dialogue can generate innovative solutions and strengthen team dynamics.

This form of dialogue requires a particular attitude, especially from the captain.

An experienced instructor explained: "The crews that perform well, and what we consistently observe in training sessions in the simulator, is their openness to communication." The instructor claims this has been an issue in aviation for some time:

> If a crew is flying at cruise level and begin planning their fuel requirements for the next flight, and the captain says, "For the next flight, I want ten tons of fuel," he essentially shuts down any opportunity for the first officer to participate in the planning. What he should be saying is something more along the line of, "Should we allocate about two tons of fuel for the next leg?" or even better, "How much fuel do you believe we should carry for the next leg?."

The point put forward was the importance of inviting communication. When seeking to explain the basics of communication in everyday life, Goffman introduced the idea and concepts of role-playing and role-taking, asserting that we can manipulate our social status by "playing" them in a way that works to our benefit. We do that if our behaviour attempts to control the impressions of ourselves and others. In Goffman's language about social life, interactio is seen as a theatre stage. In this life, in some sort of theatre, human life is reduced into a strategy of "impression management." When we construct a convincing "front" of ourselves, we keep our human counterparts "back stages" (Goffman 1974). It turns out dicastery if the cockpit becomes a "stage" on which the pilots cooperate in a Goffmanian way, performing their duties from such positions.

Creating a good working atmosphere in a cockpit and among the crew is the captain's primary responsibility. In aviation parlance, this skill is known as "good airmanship." It fosters communication where each individual actively and carefully listens to one another. The objective is to ensure that everyone in the crew feels to be a vital part of the crew as a collective unit. Honesty and respect in flight operations are the foundation of the team if each member has full support and attention. The backdrop for this shared perception of a crew is evident. In most accidents, when human error has been identified as the crucial factor, a lack of shared sensemaking has been revealed as a determining factor.

On a flight, an instructor and a captain-in-training were discussing cockpit communication. "The real challenge," the instructor said, "is how the Pilot Flying (PF) and the Pilot Monitoring interact." In contemporary aviation, there has been a noticeable shift in focus towards this dynamic. Where the second-in-command was once referred to as the Pilot Not Flying, with unclear responsibilities and limited authority, the role is now formally defined as the Pilot Monitoring (PM).

The change is not just in terminology. Traditionally, the boundaries and expectations tied to the Pilot Not Flying role were ambiguous, which sometimes led to misunderstandings or missed opportunities for correction. This ambiguity became especially pronounced when the First Officer - already in a junior position relative to the captain – was expected to monitor, speak up, and intervene when necessary. In recent years, however, cockpit procedures have evolved. The Pilot Monitoring is now expected to be actively engaged, not passive. First Officers are encouraged to speak their mind, question, and clarify - not despite their position, but because of it.[8]

In modern aviation, in Western airlines, the cornerstone of the PM's responsibilities rests in the monitoring of the PF's actions and decisions. This includes the oversight of the flight's evolution, a watchful eye on critical systems, and a keen awareness of all other salient factors influencing the flight. Should the PM discern any facets of the flight that warrant correction or enhancement, they are vested with the authority to proffer assistance to the PF. This form of support might manifest as recommendations, guidance, or the provision of supplementary information, all for the sake of the safety and operational efficiency of the flight.

In scenarios where swift and decisive action is imperative to address an emerging issue or to ensure the secure execution of flight procedures, the PM may be required to deliver clear and unambiguous directives to the PF. In moments of a potential critical situation, the PM is entrusted to intervene. This intervention may encompass the temporary assumption of control over the aircraft or the initiation of prescribed emergency procedures. It endows the pilot monitoring an active role, with the capability to provide input and ensure that any errors or deviations from established procedures are addressed.

A senior flight instructor elaborated on the need for a change in the role of Pilot Not Flying to Pilot Monitoring. "The West Air crash was a devastating proof of the lack of a systematic approach to have the PM speak up." This tragic accident, which occurred on January 8th, 2016, displayed in his mind the proof of the needed change.[9] The Swedish-registered aircraft, a Canadair CRJ 200, owned by the company West Atlantic AB, took off from Gardermoen at 22:08 and was en route to Tromsø, carrying a 4.5-tonne load of postal deliveries. According to the accident investigation report, everything was normal when the aircraft departed from Gardermoen. There were two crew members on board.

While leaving the Eastern Norway region behind, the aircraft climbed to its cruising altitude of 33,000 feet. As midnight fell, radio traffic was minimal. Inside the cockpit, the two pilots talked about everyday matters. Earlier that day, they had flown the same route, but in the opposite direction from Tromsø via Evenes. They were looking forward to a good night's rest before their upcoming flight south the following day.

Fifteen minutes past midnight, pilots and planes had passed Bodø and were following the shortest air route to Tromsø, which occasionally crossed into Swedish airspace. At this point, they were flying over the remote mountainous areas of Swedish Lapland, while still maintaining communication with Norwegian air traffic control. Inside the cockpit, the pilots had turned on the lights to begin planning their approach to Tromsø. Unfortunately, the reflection of the light made it impossible to see either the sky or the faint contours of the terrain below.

Twenty minutes past midnight, the Joint Rescue Coordination Centre in Northern Norway received an emergency message indicating that the aircraft was near Lake Ritsemjaure, approximately 25 kilometres from the Norwegian border. At this moment, the aircraft also disappeared from radar. The aircraft was flying at an altitude of 9,500 meters when something unexpected occurred. The cockpit voice recorder registered the captain exclaimed with a strong expression that the autopilot disengaged, and the aircraft started losing altitude. Sixteen seconds later, the crew transmitted a brief distress call without providing details about what had happened. Seventeen seconds later, the aircraft reached the maximum allowable speed for this type of around 580 km/h.

Just one minute and twenty seconds after the captain's notice, the aircraft crashed to the ground with tremendous force, creating a crater six meters deep and twenty meters in diameter. The last recorded speed was 508 knots, equivalent to about 940 km/h.

Precisely 19 minutes past midnight, one of the so-called Internal Reference Units (IRUs), which measures the pitch and roll angles of the aircraft, on the captain's side of the instrument panel, malfunctioned (Figure 6.1, on picture to the left). This was an exceedingly rare failure, completely bewildering the captain. Suddenly, the artificial horizon incorrectly indicated that the aircraft's nose was rising. Initially, the error was minor, but within seconds, the instrument showed an increasingly steep nose-up attitude. At the same time, the autopilot disengaged, triggering a loud alarm that filled the cockpit with intense sound.

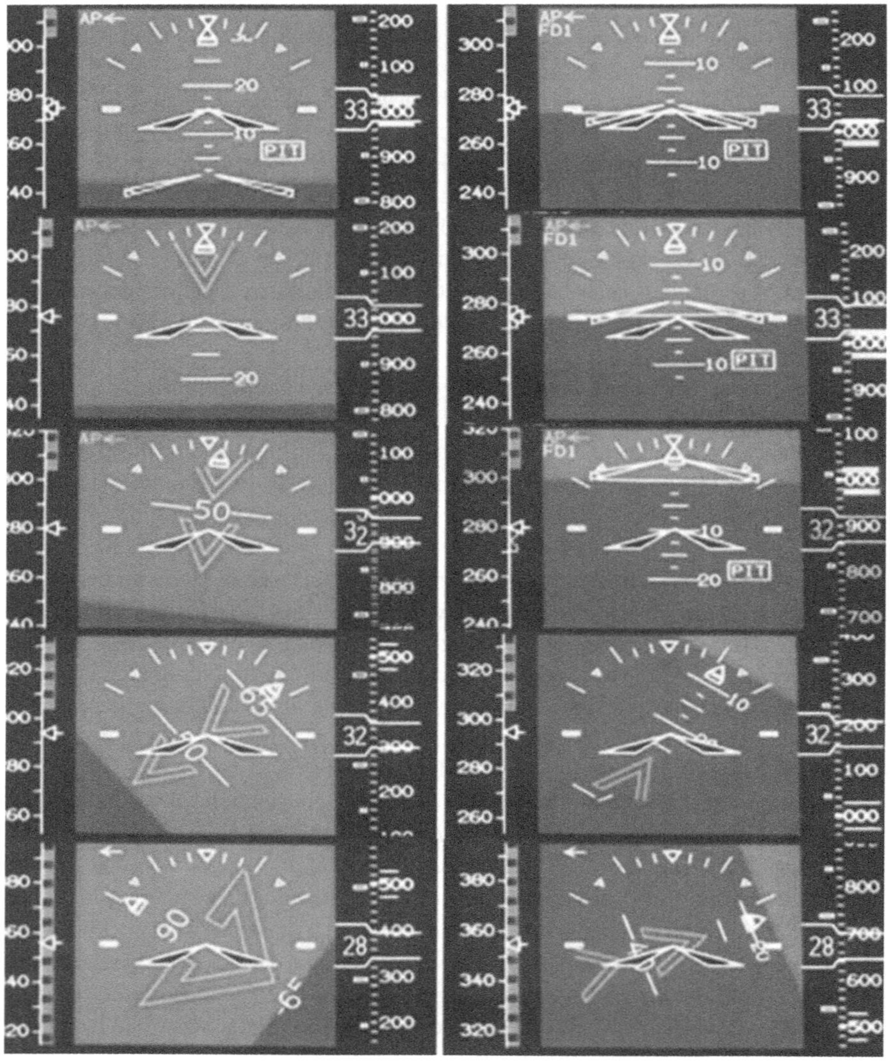

FIGURE 6.1 The difference between the left and right pilots' flying displays.

(Photo: Swedish Accident Investigation Authority (2016))

If the aircraft had responded with a rapid and steep climb, it could have entered a dangerous situation. Therefore, the captain instinctively pushed the control stick forward, even though the aircraft was, in fact, at flying level. This sudden dive manoeuvre executed at high speed left both pilots weightless, and loose items floated up towards the ceiling.

"Damn!" the captain shouted in confusion.

The first officer was taken aback by the captain's unexpected manoeuvre. On his own instrument panel, everything appeared normal, and it seemed as if the aircraft was climbing steeply. After a few seconds, he realized that it was actually in a steep descent and voiced the same expletive as the captain.

"Climb up!" shouted the first officer.

"Help me, please!" replied the captain, who still believed that the aircraft was about to stall due to a high angle of attack on his control panel, despite the input not helping.

In the growing panic, both pilots experienced tunnel vision. If the captain had taken a moment to look at the first officer's instrument display or checked the central reserve gyro, he might have discovered the error. Meanwhile, the first officer assumed the captain was struggling to pull the aircraft out of the dive. If this instrument failure had occurred in daylight, a glance out the window could have averted the catastrophe. But it was dark outside, and the aircraft had begun rolling to the left. The alarm indicating an overspeed generated an even more deafening noise in the cockpit, and the crew became increasingly disoriented.

"Turn right!" the first officer shouted, unaware that the captain's gyros indicated they should turn left. In a panic, the first officer pressed the radio button and transmitted multiple Mayday signals. The air traffic controller acknowledged receipt of the distress signals but had no option other than watching the radar screen as the aircraft descended at a speed approaching the speed of sound.

"We need to climb," the captain finally claimed, but the instruments still indicated the aircraft's nose was pitched up, and he continued to push the control stick forward.

"Help me, please!" he pleaded desperately.

The aircraft was now completely inverted, and the nose continued to point downward as it plummeted towards the ground at an irreversible speed. The last recorded seconds on the voice recorder were filled with profanities before the recording abruptly ended.

NOTES

1 The International Civil Aviation Organization (ICAO) is a specialized agency of the United Nations established in 1944. It is responsible for developing international standards and recommended practices (SARPs) for civil aviation to ensure the safe, efficient, and orderly growth of global air transport. ICAO provides a framework for member states to harmonize their aviation policies and regulations, facilitating collaboration across borders. Its key areas of focus include aviation safety, security, environmental protection, and the improvement of air navigation services.

2 CFIT, an abbreviation for Controlled Flight Into Terrain, refers to an accident where an airworthy aircraft is unintentionally flown into terrain, water, or an obstacle while under the control of the flight crew. These incidents typically occur due to pilots' lack of awareness or misjudgement of the aircraft's position relative to the ground. CFIT accidents often happen without mechanical failures, and the crew remains unaware of the imminent disaster until

it is too late. Such accidents have historically been a leading cause of aviation fatalities, prompting advancements in terrain awareness systems and pilot training programmes to reduce their occurrence.

3 https://web.archive.org/web/20210730033840/https://aviationsafety.net/investigation/cvr/transcripts/cvr_av052.php Retrieved 25 January 2025.

4 Aircraft Accident Report, Avianca, The Airline of Colombia, Boeing 707-321B, HK 2016, Fuel Exhaustion, Cove Neck, New York, January 25, 1990 (PDF). National Transportation Safety Board. April 30, 1991. NTSB/AAR-91/04. Archived (PDF) from the original on November 29, 2015. Retrieved 15 January 2025.

5 H. P. Ruffell Smith. A Simulator Study of the Interaction of Pilot Workload with Errors, Vigilance, and Decisions. NASA Technical Memorandum 78482. Ames Research Center, Moffett Field, California.

6 The cockpit authority gradient refers to the established hierarchy and distribution of authority within an aircraft's cockpit, encompassing the command and decision-making dynamics among crew members. A steep authority gradient occurs when one individual, typically the captain, holds dominant decision-making power, which can suppress input from subordinate crew members. Conversely, a shallow authority gradient represents a more balanced distribution of authority, encouraging open communication and collaborative decision-making. Maintaining an optimal authority gradient is crucial for effective crew resource management, as both excessively steep and overly flat gradients can hinder performance and compromise safety.

7 National Transportation Safety Board. (2007). Attempted Take-off from Wrong Runway, Comair Flight 5191, Bombardier CL-600-2B19, N431CA, Lexington, Kentucky, August 27, 2006. Aircraft Accident Report NTSB/AAR-07/05. Washington, DC.

8 A noteworthy example of this transformation is observed in a Scandinavian airline's adoption of the MADI framework. This innovative approach introduces Monitoring, Assistance, Direction, and Intervention as key principles, providing a systematic blueprint that clearly outlines the responsibilities and prerogatives of the Pilot Monitoring (PM). Emphasizing their active and participatory role in supporting the Pilot Flying (PF), the MADI framework represents a significant step towards enhancing communication and collaboration in the cockpit.

9 The following text is written based on the official accident investigation report. Swedish Accident Investigation Authority. 12 December 2016. Retrieved 12 December 2016. https://shk.se/download/18.2d6f089b18faca29dc81cffa/1700491384504/RL-2016_11e.pdf. I have also read several newspaper articles and online chats about the accident.

7 Beyond Communication

This chapter seeks to unravel one of the complex mysteries in the field of human factors and tries to get closer to something that often slips past us. In flight operations, communication is more than calls, checklists, and routines. At the heart of it lies something else. Something harder to grasp. It is when talk becomes interaction – when words are not just exchanged but responded to, when people tune in to one another, in the moment.

It is easy to think of cockpit communication as a technical matter. And in many ways, it is. But it is also situational. Emotional. Cultural. It carries tension and reassurance, caution, and routine. And sometimes, it carries silence. This chapter is about those moments. The ones that do not fit neatly into procedures but still shape the flight.

The experienced instructor "Per" explained:

> On a normal day when everything goes according to plan, and all systems work flawlessly, there is no apparent challenges. It is when the "shit hits the fan," and the crew needs to do some unusual while flying you will see the real good crews.
>
> *(Interview, 2020)*

The pinnacle of crew performance reveals itself when, amid the turbulence, both literal and metaphorical, the crew's ability to stick together as a crew is what prevails: "The thing that differentiates a crew that handles difficult situations effectively from those who do not, is not just about communication, but the relation between the pilots" (Interview, 2020).

Pilots' ability to interact in a good manner is not to be taken for granted. Anyone who has worked together with someone knows the implicit struggle and not just the joy, about it. It is crucial to know how to fill the gap in conflicting working relations. This universal issue must also be addressed and implemented in flight safety, where crew relations are so intensely important.

7.1 THE SMOOTH MAKING OF "WE" AND "US"

Aviation has its unique version of the matter, setting it apart from most other professional fields. Unlike many others, aviation demands the creation of a new team often, sometimes each day, sometimes on every flight. This makes a culture of rapid change that, at the same time, relies on basic continuity.

A flight crew comprises a combination of pilots and cabin crew, each possessing unique skills and responsibilities. While other professions often involve fixed teams or colleagues who work together over a long time, aviation life is embedded in a short-term framework with a high degree of change and variability. Still, the well-experienced actors are supposed to make a high degree of continuity on a short-term basis in work operations. Commercial airline operations demand flight crews that are assembled and organized for each flight, and frequent turnover extends to both the

DOI: 10.1201/9781003725862-7

cockpit and the cabin. A typical scenario in commercial aviation is that a pilot may commence the first flight of the day with a particular set of colleagues, including other pilots and cabin crew members. As the day unfolds, subsequent flights may introduce an entirely different ensemble of individuals with whom they must immediately form a functional team. This rapid and dynamic shift in team composition can occur due to various factors, including crew rotations, changes in flight routes, and the inclusion of reserve or standby personnel.

The complex of such shifting daily team formation can contain particularly challenging implications for flight safety and performance. The psychological climate within a crew, affecting collaboration and well-being, plays a crucial role in the crew's ability to perform tasks safely and effectively. A pilot may not perform at the same level in different flight crews. The same pilot may perform excellent in one crew and a poorer one in another. Just as the flight crew can influence each individual pilot, either positively or negatively, a single pilot can also have a significant impact on the overall psychological climate of the crew. The composition of a crew can change, and the performance and dynamics of the crew can vary when a new pilot replaces another. The psychological atmosphere established between crew members during a long and busy shift may change during an exhausting day.

A semi-experienced pilot working in a commercial airline described the challenge:

> Every day I go to work, it's like we start by creating a "we." It is "us" as a group who will make the day. I often think the job is a bit like playing cards. Every day, a new hand is dealt. It is up to each of us to think: Here are the cards I have been dealt today. Some days, I'm dealt good cards. Other days, not-so-good. I can't decide the cards I'm dealt. The ones I'm set up to fly with, I have no control over. But it's up to me to make sure that the crew members I've been assigned with become the resources I need them to be through the day. I need my co-pilot to be the resource I need him to be. I need the cabin chief and the flight attendants to be the resources I need them to be. This applies to everyone involved in an operation: the technician, the ground crew, and the Gate personnel. Everyone involved in the operation.
>
> *(Interview, 2010)*

> It is not like you can expect people to be "professional." I think of myself and the active part I play in people acting as good as they can be. That is why I am always trying to be polite towards my crew. I must be approachable to everyone and try to make them feel safe to speak up if they see something that is not looking right.
>
> *(Interview with "Henrik," 2020)*

Outside aviation, the terms "collaboration," "cooperation," and "interaction" are frequently used interchangeably. The concept of "collaboration" is intrinsically linked with its English counterpart, "cooperation," deriving its etymological roots from the Latin word "cooperari," signifying the act of working together. Cooperation is the common effort to solve a task or a problem that can only be solved through communal action in a group (Anderson, 2012; Karlsson & Borg, 2013; Strong, Sutherland & Ness, 2011; Roschelle & Teasley, 1995). Collaboration denotes the tangible, task-oriented aspect of the collective pursuit of well-defined tasks. Work assignments are distributed among participants, and each assumes a role of "commitment and responsibility" in executing their designated tasks, all in pursuit of a common goal (Roschelle & Teasley, 1995).

These theoretical nuances are probably more interesting for academics. However, in the world of aviation, interaction is a key term.

The verbal communication, nonverbal cues, gestures, or actions between the two pilots and the flight attendants can be cooperative, collaborative, competitive, or conflictual, depending on the nature and objectives of the people involved. However, the idea is that they interact.

The Flight crew must cooperate, and they must interact while doing so. The way the crew communicates and interacts will influence how the crew performs. In turn, it reflects how well or poorly they carry out the operation. In commercial flight operations, one of the most important tasks is to build the crew. Not just as individuals, but as one operational unit.

This team is always temporary. Crew members may not know one another well, yet they are required to establish trust, coordination, and a shared understanding within a very limited time. What makes this possible is interaction. In this context, interaction should not be understood as communication between individuals, nor as a vague form of team synergy. Rather, interaction gives rise - here and now - to a temporary shared operational space in which individual actions become mutually intelligible and practically consequential. This space is created in the moment and does not belong to any single crew member; it belongs to the crew as a whole. It exists only as long as interaction is actively sustained. I use the term shared operational space to describe this temporary, interactionally produced condition. It is within this shared operational space that procedures can be enacted as intended, mutual support becomes possible without negotiation, and the crew is able to stay ahead of the aircraft. What is often described as "crew mentality" is not a precondition for safe performance, but an effect of this interactional space being successfully established. Tony Kern, a recognized expert in the field of human performance and safety, quotes a French general to highlight the importance of developing this collective mentality: "Four brave men who do not know each other will not dare to attack a lion. Four less brave men, but knowing each other well, sure of their reliability and consequently mutual aid, will attack resolutely" (Kern, 1997).

In various cultures in the aviation industry, there is an (outspoken) recognition of the inherent interdependence among the individuals in the team. Pilots often pinpoint this argument, and illustrated in Figure 7.1.

> Being a good pilot means to belong to a good crew. And being a good crewmember may lead to being a good pilot. The successful and safe flight is not an individual pursuit but a collective endeavour, where every individual involved contributes to the overall outcome.
>
> *(Interview, 2009)*

In interviews I have conducted over the years, pilots underline that creating a team is a task in itself:

> Being good at working close with others is important. Knowing where and when to bring up a subject is cardinal. A good pilot corrects his partner when the timing is right. If things occur during take-off, you might want to wait until you've reached cruise level. Sometimes, you must wait to correct a college until you've reached ground. On a day-to-day basis, on every leg, you must feel your way to when and how you deal with

FIGURE 7.1 A good enough relationship: trust and dialogue carry the flight through cooperation in the cockpit.

(Photo: Christian Aasen Stratton).

> your fellow crew. Sometimes you must react instantly. If the situation is getting out of control, you take instant action of the situation, regardless of whether you are captain or first officer.
>
> *(Interview 2005, Nergård 2014)*

7.2 THE PROMPT INTERACTION

The short-term bonding between the pilots and the rest of the crew is crucial in the dynamics of flight operations. While their time spent together may be limited to a specific flight or duty, establishing a productive working relationship is nonetheless essential for safe operations. A captain tells what it may be like on the flight deck:

> Meeting a new pilot for the first time starts with small talk. During the briefing, we do what we must. We get the weather, look through the tech log, discuss and inform the cabin crew, and try to create a relaxed working atmosphere. The first meeting is also much about learning and understanding one another. I also think: "How will we react together if something happens? When flying with a new colleague I have to think of the unpredictable." When we sit together in the crew room and plan a landing somewhere. Through the small talk and the practical planning of the flight, it is almost as if my experience from playing in a music band comes back into play. The other pilot and I must find our "chamber tone." This chamber tone will help us to solve the foreseen as well as the unforeseen because we agree on the fundamental. We created a "tone." Sometimes

> I think creating the relationship with another pilot, and the whole crew for that matter, may be more demanding than the flight itself. It happens I return home from work very emotional tired, especially if the crew do not a find a common tone.
>
> *(Interview, 2022)*

The expression "chamber tone" originates in music, where it refers to the reference pitch to which instruments are tuned before they can play together. Its function is practical. It establishes a common point of orientation that allows different instruments to sound compatible. At the same time, it presupposes attentive listening and mutual adjustment, giving it an inherently social dimension. In the cockpit, the pilot uses this metaphor to describe the working relationship that must be established between crew members. Finding a common chamber tone is not primarily about personal chemistry or shared experience, but about agreeing on fundamental ways of working - pace, expectations, and how actions are coordinated. In this sense, the chamber tone provides both an operational reference and a social basis through which a shared operational space can emerge.

The flight community brings together professionals with diverse backgrounds. It is not just individuals who meet in the cockpit. The many cultural backgrounds that stick together there may influence flying and security extensively, and the issue has probably not been sufficiently focused on. Nobody knows for sure what impact it may have on flight operations, especially during critical circumstances. The two pilots on a flight deck might even have backgrounds that present challenges in understanding and appreciating different perspectives. Especially when faced with emergencies, the "I" and the "You" positions should merge into a collective "We."

Cultural differences in a cockpit are far more nuanced than they might initially appear. While the term "culture" often evokes national or ethnic identities, its influence extends well beyond these boundaries. Pilots bring with them not only their personal cultural backgrounds but also the imprints of their professional environments, training philosophies, and ingrained communication habits. These layers of culture – national, organizational, and interpersonal – intersect in the confined space of the cockpit, where collaboration under pressure is critical. Pilots also bring their "family tone" with them (Nergård 2005).

The cockpit is a unique microcosm where individuals with potentially divergent approaches to hierarchy, authority, and problem-solving must operate seamlessly as a team. Differences in working cultures, such as those shaped by training organizations or airline policies, can affect decision-making styles. Similarly, variations in communication culture – whether direct or indirect, assertive or deferential – can shape how crew members share critical information. When these cultural layers conflict, they may disrupt the synchronization necessary for safe and effective flight operations, especially in high-stakes situations.

The Kenya Airways Flight 507 accident provides a vivid example of how these overlapping cultural dynamics can influence cockpit interaction. The flight departed Douala in Cameroon on the night of 5 May 2007 under challenging weather conditions. It was operated by a Boeing 737-800 with a crew of six and 108 passengers. Shortly after take-off, the aircraft entered an uncommanded right bank, continued rolling, and crashed into a mangrove swamp. All 114 people on board lost their lives.

At the core of this tragedy was not just a technical failure, but a breakdown in teamwork and communication between the pilots. The final investigation, published by the Cameroon Civil Aviation Authority (2010), concluded that the crew lost control of the aircraft due to spatial disorientation. The autopilot was never properly configured and set up, and neither pilot took manual control. For nearly a minute, the aircraft banked slowly without correction. When the warning sounded, it was already too late. Both pilots attempted to regain control, but their actions were contradictory and uncoordinated. The aircraft entered an extreme bank and descended rapidly.

The captain and first officer, coming from different professional and cultural backgrounds, failed to merge their individual roles into a cohesive "we." They could not bridge their differences under pressure. The cockpit voice recorder revealed confusion, silence, and delayed responses. At no point did the crew clearly agree on who was flying. There was no confirmation of autopilot engagement, and no assertive callouts between the pilots as the aircraft drifted off course.

Instructors and colleagues had previously raised concerns about both pilots.[1] Instructors and colleagues had previously raised concerns about both pilots. The captain was described in training records and peer reports as confident and occasionally overbearing, with a tendency to dominate operational decisions. The final report notes that his behaviour may have been shaped by earlier training setbacks, and that he often assumed a dominant role in the cockpit – something not uncommon in hierarchical aviation cultures where authority is concentrated at the top. The first officer, in contrast, was described as reserved and hesitant and had previously struggled with assertive communication in monitoring roles (Cameroon Civil Aviation Authority, 2010).

I am aware that these descriptions must be treated with caution. They appear in training reports and accident records written after the event – and thus always risk being coloured by hindsight.[2]

The interactional mismatch between the pilots became evident during the take-off, as the flight faced adverse weather conditions. Thunderstorms, strong winds, and poor visibility surrounded the airport, creating a high-pressure situation that demanded sharp coordination and decision-making. Despite these conditions, the captain decided to proceed with the departure. The first officer, though likely aware of the increased risks, voiced no objection, illustrating a broader issue: the power imbalance between the two pilots stifled open communication.

Moments after take-off, the aircraft's autopilot was engaged. Unbeknownst to the crew, it was improperly configured, causing the plane to roll into an excessive right bank. Despite audible alarms warning of the dangerous angle, neither pilot reacted in time. The captain, focused on other tasks, failed to monitor the flight instruments adequately, while the first officer, seated beside him, did not intervene. The CVR captured fragmented exchanges that revealed a lack of coordination and mutual understanding. Instead of working as a team to correct the autopilot configuration, the two pilots operated in isolation, with no clear delegation of tasks or shared strategy. The "I" and "You" positions never merged into the crucial "We."

The dynamics in the cockpit of Flight 507 cannot be fully understood without considering the cultural and professional backgrounds of the crew. Studies in aviation psychology have long highlighted the role of culture in shaping cockpit interaction (Helmreich & Merritt, 2001). Pilots from hierarchical cultures may be less likely to challenge authority, even when safety is at risk, while those from more egalitarian backgrounds might expect open dialogue.

In this case, the captain's authoritative approach reflected a cultural norm where decision-making power rests with the senior pilot. The first officer, likely influenced by deference to authority - a common trait in many aviation cultures, opted to remain silent rather than risk confrontation. This silence proved fatal. Had the first officer been more assertive, or had the captain fostered a more inclusive atmosphere, the escalating banking error might have been corrected in time to save the flight.

The Kenya Airways Flight 507 accident underscores the dangers of failing to establish a shared working tone in the cockpit. The limited time pilots have to build trust and understanding becomes a significant obstacle, especially when combined with the pressures of adverse weather, technical challenges, and cultural differences. In this case, the captain's and first officer's inability to align their approaches in a high-stress situation transformed a manageable technical issue into an irrecoverable loss of control.

One pilot reflects on this dynamic:

> I sometimes notice that my approach to decision-making feels very different from my colleague's. Maybe it's about how we were trained or what we've experienced, but there are moments when I wonder if we're on the same wavelength. It's not a bad thing - it's just something I'm aware of.
>
> *(Interview, 2018)*

These subtle variations in approach can either enhance the flight operation or become a source of misunderstanding. It depends on how the relationship is managed during the limited time available.

From an ethnographic perspective, this tragedy illustrates the complex social world of the cockpit, where technical precision intertwines with human interaction. Observing the pilots' challenges in navigating their roles and communication styles reveals how deeply cultural and interpersonal dynamics can shape the outcomes of flight operations. It invites reflection on how training and preparation might better account for these nuanced, relational aspects of flying, where mutual understanding and collaboration are as critical as technical expertise. As a captain in an interview explained, creating this shared tone can be more demanding than the flight itself. "You can't play well together if you're not aligned, and that alignment doesn't happen by accident. You have to listen, adapt, and be willing to adjust." When the "I" and the "You" positions merge into a collective "We," the cockpit becomes more than just a workspace. It becomes a place where safety and trust are forged, moment by moment, through interaction.

In high-stress situations, such as emergencies, the two pilots in the cockpit must work together closely to address the problem effectively. They must rely on each other, and the breakdown in their relationship has negative consequences. In high-stress situations, the two pilots in the cockpit work together closely to address the problem effectively. They rely on each other, and a breakdown in their relationship may have fatal consequences: "In high-pressure situations, like the one they faced during the emergency, it's the two pilots in the cockpit who must collaborate closely to manage the crisis effectively," instructor "Per" commented after a simulator session (Interview, 2023). Their relationship and ability to work together were critical factors added to the Kenya Airways Flight 507 accident.

When discussing relationships among crew members, it is crucial to broaden the perspective further. Consider a typical five-day trip where the same crew operates together. It is not just the interactions within the crew room or the cockpit that matter. Every interaction among crew members plays a part. As one captain emphasized: "Even if all aspects of flying are executed flawlessly, a thoughtless comment or action during dinner after the first day can rapidly erode the trust and rapport that took a long time to build" (Interview with "Andreas" 2023). "Trust," he noted, "takes ages to establish but can be shattered in mere moments. Throughout the work period, there's a constant awareness of what's said to whom and how it's said – a continuous monitoring of one's own conduct."

In modern Western airlines, company culture and safety culture have become pivotal in shaping the dynamics between pilots. There is an increasing emphasis on fostering crew members who can collaborate effectively, recognizing that the success of a flight hinges not only on individual technical proficiency but also on the strength of interpersonal relationships in the cockpit. Historically, aviation paid little attention to these relational dynamics. Captains were primarily trained as skilled aviators, with minimal focus on the collaborative and communicative aspects of flying. This began to change with the introduction of Crew Resource Management (CRM), a framework explicitly designed to enhance communication and teamwork among flight crews. Introduced in the late 1970s, CRM not only redefined how crews interact but also ushered in a shift in company and safety cultures, embedding collaboration as a core value within airline operations.

Since CRM's establishment, airlines have increasingly adopted structured routines and practices to develop not just individual pilots but cohesive teams. These efforts reflect a growing recognition of the psychological and relational demands of the profession (Nergård et al., 2011). The quality of interaction within the cockpit is not just a matter of teamwork but a crucial factor that directly impacts the safety and success of every flight (Figure 7.2).

7.3 THE PSYCHOLOGY OF PROFESSIONAL TOGETHERNESS

The capacity for togetherness refers to the ability and willingness to engage with others in a shared professional work. It involves not only technical competence, but a relational readiness to cooperate, to adjust, and to sustain coordination under demanding conditions. From a psychological perspective, this capacity can be illuminated through the work of the British psychoanalyst Donald Winnicott. Winnicott was concerned with the individual's capacity for relatedness - that is, the ability to enter meaningful relations without losing a sense of self. He located the origins of this capacity in early relational experience, not as a deterministic blueprint, but as a set of conditions that make relatedness possible. The capacity for relatedness is according to Winnicott, rooted in the early primary relationships and works as the foundation for the personal ability to establish meaningful relations with others throughout life. The reference to

FIGURE 7.2 Creating a Crew.

(Photo Christian Aasen Stratton.)

Winnicott is not intended as a developmental explanation of professional behaviour, but as a conceptual lens for understanding the relational space in which professional cooperation becomes possible. Central to Winnicott's thinking is the concept of a "transitional" or "potential" space: an intermediate space between inner experience and external reality. Rather, it is a relational space that emerges when conditions allow individuals to engage with others and with the world in a way that is both protected and open. When people work together professionally, including in the cockpit, meaningful collaboration depends on the ability to enter such an intermediate relational space with others. Professional togetherness does not require intimacy or emotional fusion, but the capacity to participate in a shared space where attention, trust, and responsibility can be coordinated. This space must be respected as both relational and bounded, allowing individuals to engage without losing their sense of personal integrity.

In building on Donald Winnicott's work, Wilfred Bion's distinction between the "container" and the "contained" offers an additional perspective on professional relationships. Bion described how one individual or group may function as a psychological container, providing a stable space in which uncertainty, anxiety, or difficult emotions can be held and processed rather than acted out. This dynamic is particularly relevant in professional settings characterized by asymmetry, time pressure, and high stakes. Bion's concept describes how one individual or group creates a psychological space – the "container" – where difficult emotions, thoughts, or uncertainties – the "contained" can be processed. This dynamic is essential for fostering mutual trust and emotional stability in interactions.

Within the cockpit, this dynamic plays a critical role. The senior pilot, for instance, often assumes the role of the container by providing a calm and stable environment in which the co-pilot's uncertainties or stress can be expressed and held without judgement. When done effectively, this relational capacity creates an atmosphere of safety and cooperation, facilitating smooth teamwork even under pressure.

This aligns closely with Winnicott's idea of the "transitional space," a psychological realm where personal experience and external reality meet. The transitional space is where authentic collaboration and creative problem-solving can take place. For pilots, this space allows a balance between the rigor of technical procedures and the adaptability required in unforeseen situations. It represents not just a functional interaction but a meeting of minds that requires empathy, trust, and mutual respect.

The ability to empathize - to hold space for another's experience while maintaining composure - ensures that communication remains effective, particularly in high-stakes scenarios. For example, when one pilot encounters overwhelming stress, the other's ability to "contain" this stress can prevent it from escalating into tension or miscommunication. Conversely, when this capacity is lacking, emotional spillovers can jeopardize the cohesion needed for operational success. While Winnicott points out that relational capacities are rooted in formative experiences, he also emphasize that such capacities must be actively sustained. Transitional space is not a static possession, but a fragile, situational achievement that requires attention, mutual recognition, and respect for boundaries. The capacity for togetherness therefore extends beyond a general willingness to collaborate. It includes practical relational skills and competencies that enable effective cooperation: communicating clearly, coordinating decisions, and managing disagreement in ways that preserve the shared task. In aviation, pilots must not only be able to work as part of a team, but to do so with precision and reliability. This requires both relational capacity and operational skill. Winnicott's concept of potential space offers a useful lens for understanding what experienced pilots often describe in practical terms. It helps articulate a phenomenon that is difficult to define but easy to recognize for those who have spent time in a cockpit: the space between people. Not the physical distance between seats, but the relational distance that must be bridged if two individuals are to operate together. It is neither the captain's nor the first officer's space, but something shared - something that becomes possible when "I" and "You" begin to form a "We."

In aviation, this potential space is not just psychological. It is operational. The flight cannot really begin until the crew has opened this space between them. It is not about perfect chemistry or total agreement. It is about finding enough trust, enough rhythm, and enough alignment to carry the flight forward. A "good enough relationship," to borrow Winnicott's phrase. Not as something fixed, but as something that is constantly being shaped and reshaped through interaction.

This space is fragile. It does not come by itself. It has to be created, often between two people who have never flown together before. And it has to be created quickly. What matters is not that they are the same, but that they are able to meet. Only then does the cockpit begin to function as a collective.

Even when such a space is established, tensions may rise. Gregory Bateson (1972), who wrote about systems and relationships, introduced the idea of the double bind – a situation where a person receives conflicting messages at different levels of communication, making it hard to respond without in some way doing the wrong thing. It is a concept from family therapy, but it fits surprisingly well in the cockpit, too.

A first officer might sit beside a captain who says, "let me know if you see something," but whose tone, posture, or past behaviour signals that speaking up will not be welcome. The words say one thing. The atmosphere says another. And the result is hesitation. Or silence. Not because the first officer does not care, but because the space between what is said and what is meant has become too narrow, too unclear. A form of loyalty conflict arises. Speak up, and you risk tension. Stay silent and you risk safety.

This is not about personality. It is about structure. The cockpit is inherently asymmetrical. One pilot holds formal authority; the other does not. Collaboration remains possible, but it requires relational work. That work takes place in the shared space Winnicott described - through tone, timing, listening, and small adjustments. A form of micro-diplomacy enacted in real time.

Sometimes pilots describe this without using any theory at all. They just say, "It was hard to read him," or "I didn't feel invited in." And that is enough. Because in those few words lies the whole story of what went wrong – not with the aircraft, but with the space between two people.

The competence and capacity of pilots are critical for successful interaction. Competence can be improved through training and experience, while capacity encompasses inherent factors like stress management, emotional intelligence, and situational awareness. The interplay between competence and capacity is dynamic, and the ability to adapt is essential for success in aviation (Figure 7.3).

The landscape of what I would call "relational simultaneous capacity" underscores the complexity of interactions among pilots, cabin crews, and passengers. It is a dynamic and multifaceted experience where multiple perspectives and interests converge. The ability to adapt and balance these interests while considering cultural, corporate, and regulatory factors is a hallmark of effective aviation crew performance.

Group cohesion is important in teamwork. It means the capacity of a flight crew to perform effectively depends on individual capabilities as well as collective synergy. The shared psychological space in the cockpit is each pilot's capacity to manage their own emotions and accept the co-pilot. It happens when the crew's performance capacity unfolds within the group.

The teamwork reflects the broader human experience of managing relationships, emotions, and decision-making within high-stakes environments. It is a world that balances individuality and collaboration. This understanding is the foundation of the psychological dynamics involved in flying, as in life in general.

In the dynamic and high-pressure environment of aviation, the interactions between pilots are intricate and multifaceted, with numerous studies revealing the significant factors at play. The quality of a crew's performance emerges from the interplay of individual attributes and how pilots influence one another, resulting in a shared execution of their duties. An important aspect often highlighted by pilots in discussions and conversations is the dynamic nature of their interactions within the cockpit:

FIGURE 7.3 Togetherness.

(Photo Christian Aasen Stratton.)

The interactions between pilots in the cockpit are a fascinating interplay of centripetal and centrifugal forces (Stierlin 1974). These dynamics represent the forces that either draw the crew together or push them apart. Interaction is a delicate balance that shapes the cooperative or potentially discordant nature of the flight crew's relationships.

Over the years, pilots have underlined factors in their interactions that have a gravitational force that pulls crew members together, creating closeness, unity, and a strong sense of belonging. These dynamics thrive on factors that encourage emotional intimacy and a sense of unity and safety within the group. They enable the crew to adapt to challenges and stressors by relying on their internal resources and bonds. On the other hand, the factors have a binary effect and can play out as centrifugal dynamics and a disruptive force that pushes crew members apart, creating distance and discord. These dynamics often stem from conflicts, unresolved issues, or even secrets within the team, leading to emotional distancing, isolation, and strained relationships. When centrifugal forces take hold, the crew's ability to cope with difficulties weakens, contributing to the dysfunction.

The pivotal centripetal force that mutual trust between pilots plays a critical role in either enhancing or diminishing psychological cohesion. The erosion of trust can introduce division within the crew, highlighting the binary nature of factors affecting pilot interactions.

NOTES

1 The characterizations of the captain and first officer in this section are not direct conclusions from the official accident report but are based on documented training feedback and colleague testimony referenced in the report. This interpretation is informed by established research on cockpit dynamics and cultural influences on crew interaction. It reflects a theoretical reading of how interpersonal and procedural factors may have interacted in the moments leading up to the crash. The aim is not to individualise blame, but to shed light on the relational and cultural dimensions of cockpit coordination under pressure.

2 In aviation, there is a justified reluctance to psychologise individuals posthumously or attribute causality to personality traits. The industry has developed a strong culture around avoiding speculation and maintaining analytical discipline when interpreting tragic events. I share that caution. At the same time, my aim here is not to evaluate the pilots as individuals, but to understand the cockpit as a relational space. What matters is not who they were, but how they interacted – and how this interaction was shaped by wider professional, cultural, and organizational frames. As an ethnographer, I must walk a fine line: staying close to the data and the official findings but also allowing space for reflection on the human and social dynamics that unfold in critical moments.

8 Airmanship

Airmanship is a term used by pilots. For years, I have tried to find out the meaning of the term. It is not just about flying planes but about how pilots carry themselves, how they develop judgement, and how they deal with the unexpected. It is more than just skills or knowledge, but about a certain kind of mindset, a way of thinking and being.

In the cockpit, airmanship is about reading the aircraft, the sky, and oneself and others – all at once. Is about feeling vibrations, hearing the engine's hum, and sensing the movements of the plane. It is about being aware, even when things do not go as planned. It is about judgements and knowing when to act and when to hold back. Airmanship is keeping safety first.

Airmanship is also about other people. It is about how to speak, how to listen, and how to lead and be led. It is about building trust and being trustworthy, and being predictable. Airmanship is not just about handling the controls. It is more of an attitude and an intuition, and learning from experience. It is about seeing the big picture while staying on top of the details. It is about balancing confidence with humility, knowing your strengths and limits.

The term "airmanship" was not something I brought into the cockpit – it was already there. It was passed down from experienced captains to young pilots, not through manuals but through stories and shared experiences. It is something you grow into, shaped in every flight and every challenge.

Any knowledge, culture, skill, and understanding must be paired with a certain way of being to be truly effective. It is about a particular way of being – a standard of conduct that defines professionalism and integrity in practice. In aviation, it is called airmanship. In seafaring, it is called seamanship.

8.1 BEST PRACTICE AS AIRMANSHIP

Throughout history, anthropologists and psychologists have studied societies that revered individuals possessing the knowledge and skill to navigate through uncharted territories – whether by land, sea, or by air. Pilots of today do not just navigate the complexities of global airspace. They also develop and design it, and thereby occupy a distinctive role in cultural expansions worldwide. That harks back to ancient traditions of the pathfinder. In Carl Gustav Jung's framework of the human psyche, the archetype of the pathfinder resonates deeply within the "collective unconscious," reflecting a universal motif that transcends individual experience. The Jungian concept, inherited from classic philosophy, posited that the collective unconscious comprises a reservoir of memories, symbols, and motifs. Jung's concept of the archetype brings into prominence that the individual memory of history is not entirely personal but belongs to the tradition itself. Everyone inherits the stored ancestral memories that Jung mistakenly thought of as psychological and named them according to that. One of his cardinal ideas was to explain how human history sorts out what experience is important to take

 DOI: 10.1201/9781003725862-8

care of and transfer to the future human community. His idea is a great humanistic idea of history, but unfortunately, it is based on a poor opinion of culture.

Ethnographers have studied the inherited memories as the cultural patterns stored in the narratives of cultures and practices. Each of them has a narrative history that works as the storehouse for the basic experience and practices bringing human life further in any community. The difference between the two traditions is a turn from psychology to culture – from the mental and cognitive to human life practices (Nergård, 2021). The "archetype" is the basic standard and state of the art of skilled work in any community. It is "the best practices" as it has developed and accelerated in terms of the fast-growing data and computer technology. This development has followed our history and is shared across societies and cultures simply because they have been proven to be the best. We may say that the pathfinder archetype in piloting is rooted in the human instinct to explore and discover. My further use of the Jungian phrase of archetype is synonymous with "norm" and "standard" as an expression for best practice. Even if the Jungian phrase somehow has become outdated, it has at the same time almost become a part of everyday language as a general and public term for the origin of the phenomena in question.

In modern aviation, pilots embody the archetype of the pathfinder even if their role transcends mere technical proficiency. Still, it symbolizes bravery, foresight, and the pursuit of knowledge, echoing the ideals of the hero's quest for self-discovery and the integration of unconscious aspects of the psyche, as Jung points out. By interpreting meteorological data, charting precise flight paths, and maintaining composure in challenging conditions, pilots tap into this deep-seated human aspiration to explore and expand horizons. They are expected to serve as custodians of safe passage. In doing so, pilots not only fulfil their role but also embody a profound psychological satisfaction that resonates with humanity's innate desire for adventure, connection, and, in the West, exploring "the world.". Figure 8.1 is an example of an airman.

FIGURE 8.1 An airman.

(Photo: Christian Aasen Stratton.)

8.2 SOCIALIZING WITH THE ROLE

The aviation world often carries an aura of exclusivity and prestige. Learning to fly is a dream for many young kids, but it becomes a reality for only a few adults. This is not just about who can access flight training, but also who excels in the field once they are part of it. The term "right stuff," popularized by Tom Wolfe in his book describing the first U.S. astronauts and test pilots, points to the personal characteristics and skills to fly and embody certain individual qualities such as quick decision-making, calmness under pressure, and exceptional situational awareness. These pilot qualities belong to just a small group possessing a natural aptitude and an almost innate ability to master the skies. Their status resonates deeply within the pilot community and the public, creating cultural awe and admiration. Piloting symbolizes smartness and human progress, and the spirit of exploration. The advanced training and ability to control sophisticated aircraft and advanced technology elevate pilots' status. The narratives of piloting bravery, quick thinking, and calmness under pressure resonate emotionally, reinforcing their enduring symbols of heroism and leadership.

On the other side of the spectrum are those considered to be "left behind." This term from the piloting jargon refers to acting candidates in the business who, despite their passion and determination, find it challenging to reach the levels of proficiency and recognition as their "right stuff" counterparts. The reasons for that may vary, ranging from financial constraints limiting access to continuous training, personal circumstances, or even differing innate abilities and learning paces. This dynamic between the "right stuff" and the "left behind" is an elitism with profound cultural and psychological dimensions, shaping societal perceptions of heroism, talent, and leadership, not just in aviation.

The idea of "the right stuff" harks back to the heroic narratives of pilots in wartime, particularly during World War II. Winston Churchill famously boasted that the pilots who participated in the Battle of Britain had exceptional qualities: "Never in the field of human conflict was so much owed by so many to so few." The statement encapsulated the idea that pilots were not just skilled but extraordinarily so, possessing qualities that set them apart from the average person.

They were seen as the epitome of courage and determination, possessing almost superhuman strength to manage fear and perform under extreme pressure. They materialized "the right stuff."

Today, the term has evolved into "the right stuffing," reflecting a more comprehensive and humanized understanding of what makes a pilot suited for flying. This shift acknowledges that the demands on pilots have changed, especially with respect to technological advances and the socio-political landscape. The demand for modern piloting is more closely related to the old term of airmanship and being "inside" the key concepts of modern aviation: acting skills and the relevant knowledge for situational awareness and communication.

The group named "left behind" has historically been those who could not meet the acceptable standards of airmanship. Especially during times of war, these pilots were often overshadowed by the celebrated feats of the few. In peacetime, they struggled to match this highly competitive field, where the line between being competent and exceptional is subtle and subjective. The historical narrative often focuses less on the individuals, creating a skewed perception that does not necessarily reflect their value and capability.

Pilots must be trustworthy. In the eyes of the public, they are under constant assessment. Trust is not limited to technical skill or performance in the cockpit. It extends to conduct, demeanour, and judgment more broadly. There are both spoken and unspoken expectations of how a pilot should act, speak, and even dress - expectations that apply also beyond duty hours. These tacit codes shape how pilots are perceived and evaluated, and they are central to the logic of pilot selection. The underlying question is not only whether a candidate can fly an aircraft, but whether they can be trusted with what others hold dear. Are they someone we are willing to send off with responsibility for lives, safety, and confidence placed in their hands?

Across diverse cultures, the role of the pilot transcends mere profession. It is viewed as a noble calling steeped in profound respect and admiration. Just as ancient mariners relied on skilled navigators to chart courses through uncharted waters, modern societies rely on pilots' ability to interpret meteorological data, plot intricate flight paths, and maintain composure under pressure. The expectation of pilots elevates them to the status of modern pathfinders.

Historically, pioneers as Charles Lindbergh, Amelia Earhart, and the Wright brothers epitomized the spirit of exploration and adventure inherent in the Pathfinder archetype. These aviators pushed the boundaries of human achievement, demonstrating how flight could connect distant cultures and transform global communication. Their courage and innovation laid the foundation for modern aviation, inspiring generations of pilots who continue to embody their legacy of exploration and discovery. In the present day, pilots encompass a multitude of expectations that reflect their status. They are entrusted with the safety and well-being of passengers and crew, requiring a steadfast commitment to safety protocols and rigorous training in handling diverse operational scenarios. Their ability to navigate through congested airspace, adverse weather conditions, and technological complexities underscores their role as custodians of safe and efficient air travel.

Pilots are leaders within the aviation industry, responsible for making critical decisions that impact flight operations and passenger experiences. Their leadership extends beyond the cockpit operations. They are also, in ethical terms, responsible for defining the role that upholds professional standards of conduct and integrity among the crew. This ethic is embodied in the demands and commitment in decision-making and accountability in all professional duties.

From a societal perspective, pilots serve as trusted ambassadors of global connectivity and cultural exchange. They facilitate the movement of people and goods across continents, fostering economic development and enhancing international relations. The role of pilots in bridging geographical distances and cultural boundaries underscores their significance as facilitators of global mobility and understanding.

Over the years, I have developed a profound respect for the women and men in the aviation industry. The highest compliment pilots can bestow upon one another is encapsulated in the phrase: "That pilot is an airman." This saying conveys a world of deep meaning. In my early studies of pilots in the mid-90s, I did not grasp the full significance of the term, and I dedicated several years to understanding its true meaning and judgement. Through a series of articles (Nergård, 2015, 2019, 2021), I hoped to grasp the debt behind the saying so deeply rooted in aviation culture. As many of my interviews have been conducted in Scandinavia, the Norwegian phrase "hel ved" (solid wood) was often used as a follow-up description of a pilot acting with admirable airmanship. The term translates to "solid" or "reliable" in English, often used to describe someone dependable, trustworthy, and of good character, like "Being the real deal." The phrase aptly describes a person with integrity, someone you can count on. The term connotes someone who has the integrity and knowledge and is dedicated to flying. This character is proud of what he does and who she is.

Airmanship is honed through years of experience and mentorship from seasoned aviators. Integrity is indispensable, like in any other true craftsmanship (Sennett, 2008). Over the years, I have noticed the dedication evident in every aspect of true airmanship, from meticulous pre-flight preparations to thorough post-flight checks. Patience, precision, and an unwavering commitment to perfecting one's skills define the airman's work ethic. Even in the face of challenges, observed from the jump seat, truly dedicated pilots remain steadfast, learning from each experience and continually seeking improvements.

Airmanship is embedded in an improving practice of the cultural ethos of the pilot community. It involves internalizing attitudes and behavioural standards that are outspoken ideals in the pilot community. It includes learning the unspoken rules and values that govern professional conduct in aviation.

Airmanship is bound to numerous competencies that fit together in the skills required to pilot an aircraft. It is bound to advanced professionalism with a holistic approach reflecting a way of life rather than just a professional requirement. At the beginning of my studies, "airmanship" had hardly been a subject in academic studies except for appearing in the biographies of seasoned captains. The term "airmanship" had been used by the legal system applied to the professional responsibility in law and justice. From a legal point of view, the pilot and particularly the captain is responsible for the accountability of the rules and regulations governing a flight. The outcome of aviation accidents is often attributed to pilots' mistakes. For insurance claims, airlines must prove accidents to be the result of human error and not caused by organizational failure. This puts pilots in a precarious position, as left to face the legal consequences alone and caught between the airline's need to exonerate itself and the public's demand for accountability. As one captain poignantly remarked: "We are left for dead, and everyone pisses on our grave," highlighting the shallow understanding the legal system has of a pilot's immense responsibility. Pilots themselves view airmanship as a way of life.

Public and legal perspectives often fail to capture these nuances, resulting in posthumous blame that distorts the true nature of pilots' responsibilities. In the fieldwork leading up to this manuscript, it has become evident to recognize the more profound principles of airmanship and how this recognition is essential for a more accurate understanding of piloting as a systemic challenge in the creation of safety on flights.

A main finding was that rather than single elements, a universally accepted standard for good piloting is intertwined with the concept of airmanship. The pilot community does not seem to subscribe to a monolithic definition but instead emphasizes that the qualities may vary significantly depending on each specific flight operation (Nergård, 2021). For instance, skills and attributes deemed essential in military flight operations differ from those required in the commercial field. One pilot argued: "In certain military flight operations a good pilot needs very different skills and strengths than commercial pilots. Being a good pilot in military flight operations is very different from being a good commercial pilot" (Nergård, 2015). Despite this contextual diversity in operational requirements, what they have in common is the emphasis on a personal attitude and professional flying skills (Nergård, 2021).

Pilots themselves tend to emphasize the importance of internalizing a good airman's attitudes and acting standards. These standards must become deeply personal: "Airmanship is not just something you put on when you put on your uniform. Airmanship is who you are," a senior captain remarked during a seminar. This internalization process involves understanding the spoken and unspoken rules and values guiding the conduct during flight operations. Another pilot remarked, "To become a good pilot, you must have or must develop the ability to adopt the attitude of a good airman. Becoming a good pilot demand a thorough understanding of oneself, the aircraft, and the system" (Nergård, 2021). This approach to learning underscores my understanding of airmanship as a way of being rather than a professional requirement.

Pilots consistently underscore that these skills cannot be learned from books but must be developed through experience. "Being good at working close with other people is important knowledge in aviation. Daily, on every leg, you have to feel your way to when and how you deal with the other guy" (interview no. 12, 2005). Self-esteem and humility are crucial qualities for pilots and have often been highlighted by informants as essential to good airmanship. Pilots take great pride in their profession, often exuding a notable level of self-confidence. This characteristic is so prominent that it has become the subject of humour in the pilot community: "If you are in a room of 100 people and only one is a pilot, how can you find the pilot? Do not worry – he will tell you." While this self-assurance is well-known among pilots, humility is equally important for the quality of airmanship.

Modern pilot communities value individuals who are open about their limitations and errors. Although the outside world has high expectations for pilots for demanding standards, the community fosters a culture of continuous learning and mutual support. Pilot communities emphasize being humble enough to recognize their shortcomings and assist others when needed: "A good airman is humble enough to acquaint himself with his errors and limitations regarding his trade as well as the other crewmembers" (Nergård, 2021). This humility and self-awareness enable pilots to deal with complex interpersonal dynamics effectively, ensuring a balanced sense of self-efficacy and esteem.

By balancing self-esteem with humility, pilots are able to maintain the confidence required to perform their duties while remaining open to learning and improvement. This balance contributes to a safe and collaborative working environment within the crew. Pride is a recognised component of airmanship. It can be observed in how pilots carry themselves - for instance when walking through a terminal. This pride is shaped both by external recognition and by an inner satisfaction in work that is done well. It reflects a professional identity rather than a search for admiration. In my fieldwork, I have encountered this repeatedly. Many pilots live as they work. Their professional standards extend beyond the cockpit. The work is not merely performed, but owned - infused with commitment, responsibility, and a sense of purpose.

Cultural awe and admiration for pilots can be deeply understood through psychological and anthropological lenses. The pilot's role is leaning on the archetypal leader who ensures security. Humans inherently need trust and reliable leadership, especially in high-stakes situations. By virtue of their training, expertise, and ability to navigate the skies, pilots become symbols of trust and reliability. Psychologically, people feel reassured knowing that a capable pilot is in control. This role of safely leading others through potentially dangerous journeys is a core trait in the social perception of piloting, making pilots modern equivalents of ancient pathfinders. In anthropological terms, the success of such figures enhances the group's overall stability and confidence. Pilots play this role in contemporary aviation. Their ability to safely guide and protect passengers or deliver critical missions boosts societal morale and reinforces communal bonds.

The huge step in a pilot career is the transition from formal education and licensing to being hired by a professional airline. As one selection officer once remarked: "Airlines don't hire a license, they hire a future captain." The statement underscores that becoming a pilot is not about obtaining a license but about embodying the qualities and potential of a future leader in aviation.

In the text, I have highlighted the critical role pilots have in aviation safety. Pilots' perceptions of airmanship provide valuable insights into developing and assessing cockpit behaviour and safety practices: "A good airman always acts according to his level of knowledge," a captain told me. New pilots must also learn to be a part of the pilot community and cope with the burden the role carries on a personal level.

The path to becoming an airman begins with rigorous training and education. This process is meticulously designed to build a solid foundation in aerodynamics, navigation, meteorology, and aircraft systems. However, beyond these technical aspects, aspiring pilots must cultivate a mindset of continuous learning and adaptability. The dynamic nature of aviation requires that pilots stay abreast of technological advancements, regulatory changes, and emerging best practices.

Every pilot must build a solid base of knowledge. The development of airmanship can be understood as the gradual construction of such a foundation - one that provides stability while allowing for adaptation. In this sense, learning to fly involves maintaining a form of equilibrium between knowing and doing. Drawing on Piaget's notion of equilibrium in psychological development, airmanship may be seen as the capacity to remain balanced in the face of disturbance. Just as a weighted base allows an object to regain stability after being pushed, a well-developed knowledge base enables pilots to manage complexity, absorb disruption, and remain operationally stable. This learning process unfolds through the continual interplay of assimilation and accommodation. New experiences are assimilated into existing frameworks, while accommodation involves adjusting those frameworks when they no longer suffice. Learning from experience depends on this dynamic balance. It is through such adjustment that pilots develop the cognitive and emotional equilibrium required to meet novel and demanding situations. In pilot training and professional development, this adaptive process is central. Skills are not merely accumulated, but continually refined, allowing pilots to respond effectively to change while maintaining the stability necessary for safe flight operations.

Pilots develop airmanship through their conscious management of the challenges inherent to their profession. It includes the responsibility for their crew and passengers and entails setting aside personal emotions when necessary. This requirement can be emotionally taxing, highlighting the importance of airmanship and the supportive community among pilots. The articulating and communicating skills become crucial.

The expression of what we may name personal pain is intertwined with every human activity. It may be coded in very different ways of acting and behaving. Human pain is included in the common language but often has its specific cultural expressions and narratives. At times, we name pain as depression, anxiety, or even indifference. The narratives articulate and validate that such human experience is, of course, an obvious part of the flying community. It appears perhaps even more complex there in the multicultural worldwide aviation industry.

8.3 THE HUMAN PSYCHE AFFECTS FLIGHT OPERATIONS

The reader is now invited to dive into the ideas of Wilfred Bion (1897–1979) and his research and development of theories on human behaviour. He was a British psychoanalyst who provided and developed exciting ideas on the human psyche and behaviour. His ideas are, for the most part, applied in psychiatric treatment and psychotherapy, but they hold a universal perspective on human acting and behaviour highly relevant for our dealings with our common human conditions and destinations. I argue they are highly relevant to grasp life inside and outside the flight cockpit during flight operations and the way crews and passengers get along with each other.

Wilfred Bion's works on group dynamics and psychoanalysis offer basic insights into the creation of airmanship. He is best known for his theories and clinical work with patients in psychotic states and for expanding Melanie Klein's concepts of projective identification. They lived in the same period and belonged to the same psychoanalytic school.

The key concepts I will introduce for my reader are their joint venture in introducing the key concepts of projective identification (Klein) and that of container and containing (Bion). Klein's idea of projective identification describes how the mother, through attuned care, carries parts of her child's unprocessed emotions – until the child is able to contain them itself. In the mother's caring relation with her child, she serves as the supporting back-up for her child, striving to manage its life struggles, psychological and practical. The mother supports and fulfills her child's attempt to act by enrolling it into the steady fulfilment of what it wants or intends. The mother's acting works as an extension of the fulfilment of the child's intentions. The child is embedded in a training it gradually manages itself. The mother's strong identification with her child in her supporting acting forms the basis for the child's mastery of its will and intentions. It gradually manages to do what the mother is doing for it. Bion conceptualizes a related idea by the distinction between the supportive role of the container bearing what the weaker part in a relation cannot take alone (Klein 1957; Bion 1962).

The core ideas of Bion and Klein are how human interaction influences our being and human development. Their leading idea is how asymmetric human relations motivate and encourage human growth or withdrawal and decline. In the Kleinian sense, the good mother cares for her baby by bearing its pain until it is able to see why it is important to do so. The psychological growth of the child depends on its growing capacity to care for itself and become less dependent on the mother and the power connected to her authority. If this development is smooth and harmless, it supports the good development. If not, it may have the opposite result. Klein is dealing with the mother-child relation as precarious, relevant for the child developing an independent relation to the mother.

Bion argued similarly to Klein about human growth as emotional experience at its core. In a less technical language, the idea is that humans bear the burden of each other. Bion posits that the primary but unknowable reality is represented by being transformed or processed in someone's mind. The transformation process results in a product that another can observe. It might be a painting, a scientific formula, a patient statement, or, in the cockpit, an interpretation of the airspace or what the aircraft is doing in the airspace. There could be many different transformations of the same.

The original emotionally meaningful experience might be the flight itself. The pilot's mental processing of this experience results in a transformed understanding that can be communicated and shared with others through verbal and nonverbal communication or, in hindsight, through storytelling and debriefings. This transformation process ensures that the essence of the original experience is recognizable in the new form, provided the recipient has sufficient understanding, experience, and motivation.

Aviation has long focused on technical training and SOPs. But the ability to "contain" uncertainty – for oneself and for others – may be just as critical in sustaining safe operations. Bion's work helps us understand how pilots process their experiences and transform them into meaningful narratives that contribute to the collective knowledge and resilience. This transformation process is essential for maintaining the high standards of safety and professionalism that define airmanship.

Airmanship evolves through both personal development and collective learning. On a personal level, pilots must constantly refine their skills, knowledge, and attitudes. This involves a commitment to lifelong learning and an openness to new experiences and insights. Pilots must also develop emotional resilience to cope with the stresses and responsibilities of their profession. Collectively, airmanship is reinforced through the sharing of experiences and the transmission of knowledge within the pilot community. Storytelling, debriefings, and mentorship are crucial practices that facilitate this process. Experienced pilots share their experience and lessons with inexperienced pilots, helping them to handle the complexities and avoid common pitfalls.

The communal aspect of airmanship ensures that the knowledge and values of the aviation profession are preserved and passed over to the next generations in the business. This fosters a sense of mutual support among pilots.

Over time, the pilot culture and the industry have developed a tradition and safety systems that preserve the work stories. Narratives of past events, successes, and mishaps are critical in conveying meaningful learning and best practices. These narratives help new generations of pilots understand the complexity of their responsibilities and the importance of airmanship.

They promote a culture of transparency, reporting, and continuous improvement (Reason, 1997).

Airmanship is a central part of aviation culture and involves a personal and collective commitment to high professionalism, integrity, and safety standards. It is a working code shaped by experience and shared knowledge that requires pilots and other aviation personnel to assess and improve their skills and decision-making processes continually.

Aviation meets many criteria for being considered a high-reliability organization (HRO). These organizations operate under complex and risky conditions yet achieve high safety and reliability standards. In aviation, safety is the highest priority. All aspects of flight operations, from aircraft design and maintenance to pilot training and flying, are imbued with a safety-oriented approach. Aviation emphasizes continuous learning and adaptation, regularly updating procedures, technology, and training based on new information and experiences. Additionally, aviation continuously works to build resilience in its systems and operations, developing the ability to anticipate and respond to unforeseen events efficiently.

8.4 STORYTELLING AS DEBRIEFING

The tradition of storytelling serves as a culturally institutionalized practice in aviation. Unlike many other professions, the aviation community has formalized procedures to develop a pilot's competence and assist pilots and crew members with everyday social interactions and emotional management. After every flight, pilots and crew members are, in principle, legally obligated to participate in a post-operative debriefing and submit a retrospective report to the company.[1]

Crew members gather according to custom after each flight to ensure no one has experienced anything unusual. An experienced captain in a Scandinavian Air Ambulance service explained post-flight debriefings:

> After each flight, the crew gathers in the crew office. We report to the company on the crew's and the aircraft's status. After doing so, we spend a few minutes evaluating the flight. Normally, the operative debriefing is straightforward and declares that everything went according to normal operations. Occasionally, we need time to work through specific events during the flight in the cockpit or cabin. The debriefing will then continue to its conclusion with no regard to time. Sometimes, we submit a written report to our superiors because the nature of the event legally demands us to report it. In most cases, we report events that we as a crew have handled well and when we have handled a situation poorly. We hand in a written report because we feel the event had elements that other crews can learn from, helping them to be able to handle a situation similar to ours differently or better.
>
> *(Nergård, 2015)*

The primary function of the post-flight debriefing is to conduct a group-level review of events or procedures during the flight. These debriefings are institutionalized to enhance safety and individual crew members' competence (Baker, 2002; Dismukes & Smith, 2000). They serve a dual purpose: promoting cognitive reflection, as Bion (1970) described, and allowing pilots and crew members to learn from their emotional experiences. An experienced captain reflected on the importance of these debriefings, stating: "In a debriefing, you learn what and how emotions can surface and the best way to deal with them" (Nergård, 2015).

The significance of debriefings is occasionally debated (Malcolm et al., 2005; Everly et al., 2000). However, Bartone (2006) characterizes post-flight debriefings as after-action group events that can have significant therapeutic value when adequately timed and conducted with the proper focus. By positively helping pilots and crew members contextualize potentially traumatizing events, debriefings contribute to individual and collective resilience (Stueland, 2006). Through debriefing, pilots learn from their experiences as part of an ongoing learning process, acting thoughtfully rather than impulsively or avoidantly. Sharing experiences is crucial for learning within aviation.

In pilot culture, debriefings act as an institutionalized "alpha function," transforming individual emotional experiences into collective ones (Bion, 1962; Benjamin, 1991). Pilots can use existing cultural narratives, such as occurrence and accident reports, to meaningfully frame their emotions and experiences. By sharing these experiences, pilots help each other bear their burdens and learn from each other as if they had been in the same situation. An experienced pilot noted:

> When I read incident and accident reports, I try to imagine myself in the same situation. I picture myself in the same cockpit and imagine how I would react and deal with the matters at hand. When I read or hear people talking about other occurrences and accidents, I always question whether I would have done the same thing as the pilot and crew in question or whether I would have reacted differently, knowing my level of skills and competence.
>
> *(Nergård, 2015)*

Post-flight debriefings allow crew members to "let off steam" together, particularly after difficult or stressful experiences during a flight. Sometimes, these sessions are light-hearted, allowing crew members to exchange small stories and share a laugh. However, debriefing immediately after landing fulfils legal obligations and creates an emotional experience that encourages thoughtful action rather than impulsive responses during future flights. Reflecting on and verbalizing feelings allows pilots to take responsibility for their emotions and actions.

Tolerance and the ability to bear emotional pain require courage (Lear, 2009) and emotional support through others' projective identification (Greenberg & Mitchell, 1983). Without these supports, pilots might resort to defence mechanisms like omnipotence, which can hinder individual thinking and development. By institutionalizing the practice of post-flight debriefings, the aviation community fosters a culture of continuous learning and emotional resilience, ensuring that pilots and crew members are better equipped to handle the complexities of their profession.

Local knowledge is essential in demanding environments like the Arctic, where flying may require specific technical skill and understanding of the landscape, weather patterns, and operational challenges unique to the region. Tromsø Airport (ENTC)[2] in Northern Norway illustrates this well. Nestled on an island surrounded by rugged mountains, fjords, and vast tundra, the airport presents navigational challenges year-round. In winter, these difficulties intensify as extreme winds, icing, and sudden weather shifts reduce visibility and complicate landings. Recognizing these seasonal hazards, the airport's classification changes in winter to reflect the increased complexity of operations.[3] Pilots flying into Tromsø must rely on more than instruments and training – they need a deep familiarity with the region and the ability to read its ever-changing conditions.[4]

Operating in the Arctic requires pilots to develop a detailed understanding of the region's distinctive geographical and meteorological conditions. Flights often take place near steep mountain ranges, where visibility may be limited and weather conditions can change rapidly. Flying in this environment is therefore not simply a matter of navigation or route planning. It demands continuous anticipation of turbulence, shifting winds, and localized weather phenomena that can affect aircraft performance and flight safety. The pilot's task is to remain ahead of these changes, adjusting decisions and actions as conditions evolve

Even though modern advancements like improved airport infrastructure, high-tech visual aids, satellite navigation, and turbulence alert systems have made Arctic flying safer, certain airports, like Tromsø Airport, still pose considerable challenges. Winter operations are particularly demanding. The airport's single runway is frequently covered in snow and ice, requiring pilots to understand snow-clearing processes and rely on systems that measure how slippery the runway is. Additionally, the airport layout, including taxiways and parking areas, can be tricky to navigate under the region's frequent low-visibility conditions. This means pilots must adapt their strategies for take-off, landing, and taxiing to ensure safe operations in these harsh and unpredictable Arctic environments.

Approaching and departing in Arctic Airport involves navigating intricate routes influenced by the surrounding mountains and fjords. The approach often requires complex manoeuvres to avoid terrain and align with the runway under challenging weather conditions. Departures demand steep climbs to clear nearby peaks and careful navigation to avoid turbulence-prone areas. Local knowledge includes familiarity with alternative airports and diversion procedures, essential in a region where the weather can rapidly deteriorate.

Operating in the Arctic also involves navigating cultural and regulatory differences. While aligned with European standards, Norway's aviation regulations incorporate specific adaptations for Arctic operations, such as stringent requirements for cold weather procedures, de-icing protocols, and emergency preparedness. Pilots must respect local cultural norms, including environmental conservation and the traditions of the indigenous Sámi people. This is particularly important for regional airlines and air ambulance services.

The stories young pilots are told when dealing with the experienced carry a wealth of historical knowledge intertwined with the practical experiences each young pilot is heading to in flying, taking off and landing at various airports, and navigating in different airspaces. This is local knowledge made into personal experience. It merges with the knowledge preserved in these stories. A skilled pilot knows precisely the challenges of each specific approach to various airports, the intricacies of each air corridor, and the locations of weather fronts in the airspace, understanding how these factors can influence the current flight, approach, and landing at a particular airport. This knowledge belongs to the familiar with an airspace gathered in stories that have been preserved through generations of aviators, including accounts of times when things went wrong or nearly did.

These stories are essential to a pilot's knowledge (Nergård, 2019). They detail dangerous segments along flight routes, specific landing conditions, favourable and unfavourable weather conditions, and what makes a particular airspace safe to fly in during certain weather situations, seasons, or times of the day and challenging in others. The experiences once shared by older pilots with novices are passed down to the next generation when these pilots add their own experiences. Each generation inherits these stories, blending them with the personal experiences they gain in the airspace and with their aircraft.

For young pilots, this merging takes the form of a way of life – having an internalized map of flight routes and airspace (Nergård, 2019).

Experienced captains are important storytellers in the aviation community. They tell about their own experiences and those shared by others, making them relatable and valuable to novice pilots. This process is crucial for several reasons. Being a pilot is an immense responsibility that entails the technical skills of flying an aircraft and the profound duty of ensuring the safety and well-being of passengers and crew. This responsibility can be heavy, sometimes accompanied by significant stress and emotional strain. Pilots occupy a unique position in society, entrusted with lives and expected to embody trustworthiness and reliability.

As people in general also pilots and air crew suppress or conceal certain emotions. In the public eye, they are expected to maintain a calm and composed demeanour, regardless of the internal pressures they might be experiencing. This expectation includes the emotional and psychological challenges they face, as well as during near-miss incidents or the stress of making critical decisions under pressure. In these cases, crews are hiding how they are affected, but the experience itself is taken care of as a lesson to learn from. This professional calm is essential for the professional being.

To bridge the gap between the unspoken emotional toll and the need for continuous learning, storytelling within the flying community becomes crucial. Experienced captains and seasoned pilots often share their personal stories – accounts of challenging flights, difficult decisions, and moments of high stress – with newer pilots. These narratives offer invaluable insights beyond what is typically covered in textbooks or training manuals. They provide a realistic portrayal of the complexities of flying, highlighting both the technical and emotional aspects of the job.

Through these stories, new pilots learn how to handle situations that might need to be thoroughly addressed in formal training. They understand how to manage their emotions, maintain composure in high-stress scenarios, and make sound decisions when faced with unexpected challenges. The stories also foster a sense of collegial support, remind pilots that they are not left alone with their harsh experiences that many others have met and overcome.

In essence, while the public might see only the composed exterior of pilots behind the scenes, there is a rich tapestry of shared experiences and stories that contribute to the shared experience that has been turned into wisdom and resilience in the aviation community. Just as the German philosopher Walter Benjamin (1991) describes the storytellers' processing and sharing of experience in a way that can be communicated to others, captains share their flight experiences, including successes and failures, in a manner that students can understand and learn from. This helps bridge the gap between theoretical knowledge and its practical application in flight operations.

Pilots contribute by their storytelling to a collective memory in the aviation community by sharing their experiences and stories containing it. Much like the ordinary cultural stories described above, the collective memory of everyday matters of flying helps standardizes responses both to everyday situations and the almost disastrous incidents. It also ensures that critical lessons are not forgotten and can be passed down through new generations of pilots. Benjamin talks about the community of listeners that forms when stories are told. In aviation, a similar community exists among pilots. Novice pilots learn not just from their instructors but from their peers by listening to and sharing experiences. This practice has formed a supportive aviation environment where knowledge is stored and shared.

Just as each ordinary human culture stores its collective memory and identity, the stories shared in aviation communities form and transfer its state of being to the next generation of pilots, ensuring that all its basic experiences survive in lessons told and passed over to secure the culture of safety and professionalism for the future.

NOTES

1 Several regulations mandate this requirement, including EU OPS 1.085b, EU OPS 1.037(a)(2), Appendix 1 to EU OPS 1.1045 Operations Manual contents, and Annex 6 to the Convention on International Civil Aviation, Part I, paragraphs 3.2.3 and 3.2.4 (Commission Regulation, 2008). EASA's Acceptable Means of Compliance (AMC) and Guidance Material (GM) related to Part-ORO (Organisation Requirements for Air Operations) underscore the role of debriefing in enhancing flight safety. EASA's Flight Examiner Manual 2021.

2 The International Civil Aviation Organization (ICAO) assigns four-letter airport codes used for flight operations, air traffic control, and aviation logistics. Unlike the three-letter IATA codes, which are commonly used for passenger travel and airline ticketing, ICAO codes are primarily used in flight planning, navigation, and regulatory documentation. The ICAO coding system follows a regional structure, where the first one or two letters represent a country or geographic area, while the remaining letters specify the airport itself. For example, Tromsø Airport is designated as ENTC under the ICAO system. The prefix "EN" is assigned to airports in Norway, while "TC" uniquely identifies Tromsø. In contrast, the airport's IATA code is TOS, which is the identifier most commonly seen on airline tickets and baggage tags. ICAO codes provide a standardized and globally recognized system that ensures clear communication in aviation operations, particularly for flight crews and air traffic management.

3 Airports are categorized based on their operational complexity and the level of expertise required for safe operations. Category B airports, like Tromsø in summer, involve moderate

challenges such as non-standard approaches, complex terrain, or specific weather conditions. Pilots flying to Category B airports must receive additional briefings and be familiar with specific procedures, but do not require dedicated training beyond standard commercial operations. In winter, Tromsø is reclassified as a Category C airport, requiring specialized training and authorization due to heightened operational risks. The airport's location on an island surrounded by mountains and fjords demands precise navigation and approach planning. While this presents challenges year-round, winter conditions amplify these difficulties. The combination of shifting winds, turbulence, and reduced visibility makes final approaches particularly demanding. With only one main runway, pilots have fewer alternatives if conditions deteriorate suddenly. Snow and ice accumulation, both on the aircraft and the runway, introduce further risks, increasing stopping distances and requiring continuous de-icing operations. Rapid weather changes are common, with heavy snowfall and sudden whiteouts frequently forcing last-minute operational adjustments. Go-arounds - aborted landings due to unstable approaches – become more frequent, and pilots must be prepared for the possibility of diverting to an alternate airport if conditions make a landing unsafe. Because of these complexities, pilots must undergo specific simulator training or supervised landings with an experienced captain before being cleared to operate at a Category C airport. Some airlines have even developed their own training portfolio, a type of "rating," that addresses challenges for operating certain airfields – like Tromsø in "winter operations.".

4 One of the notable features of Tromsø Airport is its steep approach angle of 4 degrees, compared to the standard 3-degree glideslope found at most airports. While this difference might seem small, it makes a significant impact on how pilots must handle the approach. A 4-degree angle means descending at a noticeably steeper rate, which requires pilots to be more precise with their airspeed and descent control. Imagine driving down a hill that is steeper than usual – it requires more focus and careful braking to stay in control. Combine this with Tromsø's common Arctic weather conditions, such as snow, ice, and low visibility, and it becomes clear why this airport is considered challenging even for experienced pilots.

9 Aviation Education

Jakob Meløe, the Norwegian philosopher (1927–2023), once recounted of a conversation with an experienced sailor. When asked what it takes to learn to sail, the sailor replied: "You learn it in just a couple of hours, but it takes a long life to be good." The same holds true for becoming a pilot. Learning to control an aircraft, to manipulate the yoke and read the instruments, can be taught in a relatively short time, though not in just a couple of hours. But mastering the craft – developing the judgment, awareness, and knowledge that define an aviator – takes years of experience, exposure, and reflection.

The knowledge of flying is not just an individual journey. It is deeply embedded in a long history of aviation, shaped by traditions, best practices, and generations of pilots who have passed down their insights. Student pilots do not merely learn procedures. They step into a profession where knowledge is shared, refined, and preserved. They do not just acquire skills but inherit a way of thinking, a way of seeing, and a responsibility that extends beyond their own flight path. Learning to become a pilot is an initiation into an ongoing lineage of experience, where technical mastery is only one part of a much larger process (Nergård, 1984).

Like the sailor who learns the basic sailing skill in a day but spends a lifetime to be good at it, a pilot's education extends far beyond formal training. It is a process of atonement – learning to read the air, to sense changes before they appear on the instruments, and to understand the subtle interplay between humans and machine. It is about knowing the limits of human perception, recognizing when to trust experience and when to defer to training. Learning to become a pilot is as much about entering a structured environment as it is about mastering technical skills. Safety, precision, and an ingrained operational ethos shape how students come to see, interpret, and act within this professional universe.

One does not become a pilot by earning a certificate. The process is gradual. It cannot be taught just in a classroom. The transition from student to pilot is not just a matter of gaining certain skills. It also consists of a transformation where the individual steps lead into a shared identity, shaped by the knowledge of those who came before.

9.1 THE EVOLUTION OF PILOT TRAINING

Aviation education never follows a single, straightforward path. It grew out of different needs, shaped by the demands of both military and commercial aviation. In the early days, training was informal. Pilots learned from mentors and by personal experience, with no standardized curriculum or formal structure. It was learning by doing, often with trial and error as a calculated factor.

Military aviation was the first to systematize pilot training. During World War I, the need for large-scale, efficient pilot training became urgent. Military forces

DOI: 10.1201/9781003725862-9

realized that survival in combat required more than just flying skills. It demanded discipline, precision, and the ability to operate under extreme pressure. This led to the development of structured training programmes with clear standards and procedures. Scholars like Brady (2000) have pointed out how military aviation laid the groundwork for systematic pilot education. But this was just one piece of a much larger puzzle.

Meanwhile, commercial aviation was emerging with its own set of needs. In the 1920s and 1930s, airlines like Pan Am and KLM started developing their own training programmes, focusing on long-haul navigation, passenger safety, and operational efficiency. These programmes were different from military methods, designed not for combat but for the growing demands of commercial air travel. After World War II, civil aviation expanded rapidly. With more people flying and new international routes opening up, the need for standardization became clear. The International Civil Aviation Organization (ICAO) was established as a global authority to create common training standards. This allowed pilots from different countries to operate within a shared regulatory framework, ensuring consistency and safety across international airspace.

While military training emphasized combat readiness and survival, civil aviation focused on teamwork, communication, and safety. This difference shaped the way pilots were educated. In civil aviation, pilots had to learn to work as part of a multi-crew team, manage human factors, and navigate complex airspaces. It was about being reliable, predictable, and safe, not just skilled.

Over time, the way pilots were trained continued to change, driven by the economic realities of the aviation industry. Airlines needed a steady supply of qualified pilots to keep up with growing demand. Training had to be efficient, scalable, and standardized across different countries. It was no longer just about learning to fly. It was about producing pilots who were ready to meet the needs of airlines in a competitive global market. This shift was driven not only by safety concerns but also by economic pressures. Airlines needed pilots who could step into the cockpit and perform consistently, no matter where they were trained.

To manage this need for standardization, regulatory bodies like the European Union Aviation Safety Agency (EASA) and the Federal Aviation Administration (FAA) developed strict guidelines for pilot training worldwide. These standards cover everything from flight hours and simulator training to instructor qualifications and theoretical coursework. They ensure that every pilot meets the same high standards, regardless of where they train. But these regulations do more than just ensure safety. They control the flow of new pilots entering the industry, maintaining a balance between supply and demand.

Flight schools operate within this highly regulated system, following detailed training manuals, operational procedures, and assessment methods. These are constantly updated to meet regulatory requirements. But despite this structure, pilot education is not a fixed system. It is a living practice, constantly evolving with new technology, changing industry needs, and the ongoing tension between tradition and innovation.

One of the biggest changes in recent years has been the shift from fixed curricula to competency-based training and assessment (CBTA). In the past, pilots were trained based on a set number of flight hours and courses. But this system faced growing criticism because experience alone does not guarantee competence. Flying is not just about putting in the hours. It is about performance and decision-making under pressure.

CBTA is about measuring skill, not time. It focuses on demonstrated proficiency rather than the number of hours spent in a classroom or cockpit. Flight instructors no longer evaluate students just by counting flight hours. Instead, they assess how well students can handle the aircraft in different situations. This shift allows for more individualized learning. Some students master certain skills quickly, while others need more practice. Instructors adjust the training to each student's pace, ensuring they only move forward when they have truly mastered each skill.

The recent changes in pilot education are founded on the idea that, beyond the technical aspects of flying, the "sailor knots" in Meløe's example, the most important aspect of pilot education is to develop a mindset where pilots are always learning, always adapting, and always making informed decisions. Aviation is always evolving. It adapts to new technologies, industry demands, and regulatory changes. It balances tradition and innovation, keeping what works while constantly searching for better ways to train the next generation of pilots. Hence, the pilot education must initiate this ever-ongoing process in the student. Pilot education must establish this relationship between the pilot student and the body of knowledge of the evolving culture of piloting. Instructors play a crucial role in this transformation. Their success is not measured by how many students pass an exam but by how well those students perform in real-world aviation. The ultimate test of an instructor's work is when airlines trust their students enough to hire them.

Pilot education has come a long way from its informal beginnings. In a deeper sense, the crux of pilot education is about creating thoughtful practitioners who are ready not just to meet licensing standards but to excel in the profession and within the practice of aviation. To fully understand it, we must look beyond the curriculum.

9.2 SELECTION AS A RITE OF PASSAGE

As an outsider entering the world of aviation, the mystery surrounding its selection process is palpable. It is whispered about in hallways, analysed in online forums, and speculated upon by those waiting at the edge, hoping to get in. It carries a reputation of being rigid and unyielding, spoken of with a mix of respect and fear. Those who pass through it carry a quiet pride, a sense of belonging earned through trial. They often speak of it with nostalgia, as if remembering a rite of passage that transformed them. For them, selection is a badge of honour and a validation of their place in the cockpit.

For those who do not make it, the story is different. Selection is harsh and unforgiving, a door that closes too soon, a dream cut short. They remember it with frustration and disbelief, questioning its fairness and logic. To them, the process feels arbitrary and even cruel, an enigma that keeps them on the outside looking in.

The truth lies somewhere in between. Selection in aviation is both myth and reality, both scientific and intuitive. It is a gatekeeper, an initiation, and a search for something beyond skill or knowledge. It is about finding those who have what it takes and separating them from those who do not. It is about identifying the elusive "it" factor – something that cannot be precisely defined but is unmistakable when encountered. The selection process is about finding those who can operate an aircraft and also embody the role of a pilot.

The selection process appears as a complex system of tests, interviews, and evaluations. It is often wrapped in technical jargon and formal procedures that seem to obscure the criteria being used. But at its core, selection is about making a choice. It is about deciding who has the potential to grow into the role of a pilot and who does not. It is about finding those who can learn and adapt, who can develop the discipline and mindset required in aviation, and who can handle the demands of flight with confidence.

Aviation selection has two main stages. It begins at flight schools and academies, where candidates are assessed to see if they have the potential, abilities, and foundational skills to become a pilot. This phase is about identifying the raw materials - the cognitive abilities, motor skills, and personality traits that indicate a candidate can learn to fly. It is about spotting those who can develop spatial awareness, keep calm under pressure, and make quick, accurate decisions. This is about potential, about seeing what could be (Martinussen & Hunter 2017).

But passing flight school is not the end. After training, candidates face another test. The airline selection. Here, the focus changes. It is no longer about potential; it is about readiness. Airlines do not want someone who might become a pilot someday. They need those who are already one. This selection process evaluates whether the candidate meets the airline's standards, whether they show maturity and responsibility, and whether they fit the airline's culture and values. It goes beyond technical skills and asks if the pilot can work well with crews and ground control, represent the airline's brand, and maintain high standards of safety and service.

Although they serve different purposes, these two selection processes are linked. They are two sides of the same coin. Flight schools look for candidates who show potential and who can be trained and moulded through experience. Airlines, on the other hand, look for fully trained professionals who can contribute immediately to their operations. One is about promise and growth. The other is about performance and readiness.

This duality reflects the nature of aviation itself. It is both art and science, intuition and discipline. It requires individuals who are precise and adaptable, cautious and confident. The selection process is about crafting a pilot - not just someone who can fly, but someone who can take on the responsibility of the role.

Selection is interesting because it is part of a culture that constantly monitors the individual. It is not merely about passing a test but about fitting into a profession that demands continuous evaluation and adaptation. In aviation, selection never truly ends. From the first day of training to the captain's seat, pilots are constantly observed and measured against standards that leave no room for error. This continuous scrutiny shapes how they fly, how they think, and how they live. Selection is about discovering those who can lead, inspire trust, and carry the responsibility that comes with flying an aircraft. It is about shaping individuals into airmen.

9.3 COCKPIT AND CLASSROOM

Pilot education seems like a world of its own, separate from the usual paths of academia. It is shaped not by theory but by the demands of the profession and the aviation industry. In most professions, knowledge is built step by step, with theory coming first and practice later. In aviation, it is the other way around. Pilots learn by doing. Knowledge is not just something they understand. It is something they perform. Figure 9.1 illustrates this more than anything.

This is what makes aviation education unique. It is about mastering practical skills to the point of precision. Every pilot, no matter where they are trained, must meet the same strict requirements. There is no room for personal interpretation. Training is repetitive and detailed, designed to produce pilots who can perform procedures flawlessly. It is not about open-ended questions or debates. It is about exact execution, every single time.

Aviation education is changing, but it remains practical at its core. Universities now offer degrees in aviation, blending traditional flight training with research and theoretical studies. This change reflects a growing awareness that pilots need more than technical skills. They need to approach aviation as a complex system of technology and safety.

Not all countries follow this model. In some places, pilot training remains strictly within independent flight schools, regulated by national authorities. In others, like Norway, the United Kingdom, France, Germany, and China, aviation has been integrated into the university system. Here, students are trained not just as pilots but as

FIGURE 9.1 Training in the simulator. Here, mistakes can be made and repeated until they vanish into muscle memory.

(Photo: Christian Aasen Stratton.)

aviation professionals, learning about safety culture, operational challenges, and emerging technologies.

Despite these differences, aviation education is globally standardized. Pilot training must meet international requirements, ensuring that pilots can fly safely anywhere in the world. This shared framework creates a global profession, where every pilot is trained to the same high standards, regardless of where they learn.

Yet aviation education does not fit neatly into the academic world. Most university programmes follow a linear model – students learn theory first and then apply it. Pilot training is different. From day one, students must combine theory and practice, learning and doing, in a continuous cycle of experience and assessment.

At UiT The Arctic University of Norway, this duality is particularly evident. The pilot education programme operates between two worlds: the academic, at the university, and the operational, in training. The two do not always align. Universities value critical thinking, research, and theoretical exploration, assessed through essays, discussions, and independent study. Aviation training values standardization, procedural accuracy, and flawless execution. In the cockpit, knowledge is not something you debate. It is something you demonstrate by acting.

For students, this creates a challenge. In university courses, learning is gradual. Students build arguments, develop ideas, and deepen their understanding over time. In flight training, progress is black and white. You either pass the check ride, or you don't. You either land the aircraft safely or you try again until you can. There is no room for almost getting it right.

This constant shift between two ways of thinking – analysing and reflecting in the classroom, executing and performing in the cockpit – is challenging. As one third-year student put it, "When I'm flying, I don't have time to analyse. I just have to act. But when I'm writing my bachelor's thesis, I'm supposed to step back and reflect. It's like training two different muscles at the same time" (Interview with pilot student 2022). Of course, this is more complicated. The practical deed of flying is a continuous analysis, but it is carried out as a part of the acting and not separated from it, as you do in a university seminar.

Instructors face this challenge too. Flight instructors come from the aviation industry, trained to teach in the cockpit. They focus on procedural accuracy and performance. Academic lecturers, on the other hand, are trained to encourage discussion, inquiry, and critical thinking. In academia, questioning is the goal. In aviation, precision is mandatory.

Even time is different. Universities run on fixed schedules – semesters, coursework, and exams. Flight training does not follow this pattern. It depends on weather conditions, aircraft availability, and regulatory check rides. A student might be ready for a crucial flight test, but if the weather turns, that check ride is delayed. This disrupts the academic rhythm, where courses build on each other in a set sequence. In flying, learning is not always linear. It must be flexible.

Despite these challenges, the integration of aviation education into universities is changing how we think about pilot training. It is no longer just about meeting licensing standards. It is about developing aviation professionals who can adapt to technological change, contribute to safety research, and shape the future of aviation knowledge.

Aviation is always evolving, and so is its education. The question is not just how pilots are trained but how they continue to learn, how they pass on knowledge, refine their judgment, and contribute to a profession that spans generations. Pilot education is about shaping a mindset.

Adopting aviation into university programmes challenges teaching and learning traditions in aviation and education. The mindset of a future pilot is not shaped in lecture halls alone. Nor is it shaped solely in the cockpit. It is crafted in the duality between them – formed through theoretical training and refined through repetitive drills, checklists, simulations, and hands-on practice. It is through this blend of experiences that knowledge becomes instinct, and skill becomes second nature.

Pilot education is about more than meeting licensing standards. It is about developing thoughtful practitioners. To truly understand pilot education, we must now step into this world of practice, where theories are tested, procedures are rehearsed, and the art of flying is mastered through doing. It is in this realm of application that pilots are truly made, where knowledge and skill are fused into the instinct and judgment required to command an aircraft.

9.4 THE PILOT EDUCATION AT UIT

After the selection process, pilot education at UiT begins. Unlike private flight schools, where high tuition fees are a barrier to entry, UiT offers free education, reflecting a democratic value that all who qualify should have the opportunity to pursue aviation. This principle aligns with Norway's need for pilots, not just for commercial aviation but for critical services like the air ambulance, which is the backbone of healthcare in a country with vast distances and remote communities. The journey to becoming a pilot at UiT follows a structured progression from foundational knowledge to complex, real-world flight operations.[1] It is designed to develop not only technical proficiency but also the ability to integrate decision-making, adaptability, and situational interpretation under dynamic conditions. Each phase of training builds upon the last, reinforcing the process of transformation from student to aviation professional.

The first year on the Tromsø programme is dedicated to developing a strong academic and theoretical foundation. Students begin with Examen Philosophicum (Ex. phil.), mathematics, and physics – subjects essential to the scientific basis of pilot education. At first, these disciplines may seem distant from the cockpit, but they are fundamental in pilot education. The autumn semester consists of intensive classroom activity, where students engage in core aviation subjects such as air law, principles of flight, and aircraft structures, operational procedures, etc. Teachers try to bridge the gap between abstract physics and the real forces acting on an aircraft in motion.

As students enter the spring semester, they begin work on the ATPL (Airline Transport Pilot Licence) theoretical curriculum. This phase of training covers core areas of professional aviation knowledge, including meteorology, navigation, radio communication, aircraft performance, human factors, and aviation safety. Through this coursework, students develop an understanding of how these domains are interrelated and how they converge in real-world flight operations. The emphasis is not on isolated subjects, but on integrating theoretical knowledge into a coherent operational perspective.

Navigation is a key component of pilot training, and students begin learning general navigation concepts, the principles of earth-mapping, and how it works in aviation. Students study map projections, time zones, compass variation, and the Coriolis effect, all of which influence route planning.

They are then introduced to radio navigation, how to operate VOR (VHF Omnidirectional Range, NDB (Non-directional Beacon), GPS, and ILS (Instrument Landing System). These tools are essential for situational awareness, enabling pilots to position themselves in airspace with precision.

As the first academic year runs to an end, students make a defining transition. From May to August, they step into the cockpit for the first time, beginning basic visual flight training (VFR). This is the moment when students move from the theoretical part of the study to active flying. Their first solo flight is a milestone, marking their initial step towards professional competency. Alone in the cockpit, they are responsible for action from take-off to landing.

In the first lessons in the airplane, student pilots are taught basic skills for controlling the plane while it is still on the ground. They learn important safety steps, checking everything is working and keeping a careful eye out, moving at the right speed and direction, and how to make turns in a narrow space. The feeling of how wind affects the movements of the plane and how different ground surfaces make a difference. Then, how to park the plane correctly and check all the instruments before the next taking off is basic. A flight instructor commented:

> We have a skill standard set for the initial lessons. The students must show that they master something as easy as safely enter and exit the aircraft, coupled with proficiency in operating elevator, aileron, rudder, trim, engine controls, and flaps accurately. I expect my students to demonstrate some form of adeptness in executing correct and safe taxiing procedures under various conditions. It might seem strange for the outsider, but in my early days of instruction, I taught at a flight school in the Arctic. Learning how to taxi on an icy runway in the Arctic proved challenging for many students.
>
> *(Interview, 2024)*

A pilot student described the initial flight training:

> Flight training is complex and very detailed. After learning the aircraft and how to taxi, we are taught how to take-off and depart from the airfield, how to safely and correctly take off and leave the control zone. First, we go through all the checks we need to do before taking off and then we learn how to take off and climb to our cruising altitude. We then proceed to learn how to make medium turns while flying straight, climbing, or descending. We're taught how to keep an eye out for other planes and how to fix any mistakes we might make while turning. We practice turning onto specific landmarks and headings as well as joining the traffic pattern at the airport.
>
> *(Interview with "Irene" 2024)*

One of the students in the programme at UiT described his first flying lesson:

> The anticipation was palpable as I approached the airstrip for my first flying lesson. A few days earlier, we had had a so-called long brief, where the flight instructor took me and some of my fellow students at the University flight program and did a walk-around of the aircraft. We even took the aircraft out of the hangar, where the flight instructor let each one of us taxi the aircraft.

Taxiing an aircraft is like driving a car, but instead of using a steering wheel, you use foot pedals and hand controls. When a plane is taxiing, it moves slowly, usually on a runway or taxiway, either towards the runway for take-off or a parking area after landing. The pilot uses the aeroplane's engines and brakes to control its speed and direction, like how a driver uses the gas pedal and brakes in a car. They also use the plane's rudder pedals to steer, just like how a driver turns the steering wheel in a car:

> But this was the day. My heart raced with excitement and nervousness, knowing that I was about to take to the skies for the first time. As I met my flight instructor, a seasoned pilot, with a reassuring smile, any apprehension I felt began dissipating.
>
> We started with a thorough pre-flight briefing, reviewing the basics of aircraft controls, safety procedures, and what to expect during our time in the air. Despite my eagerness, I listened intently, absorbing every piece of information like a sponge.
>
> Once ready to go, we made our way to the waiting aircraft. A sleek two-seater trainer plane that seemed to beckon me toward adventure. After performing the necessary checks and strapping ourselves in, my instructor taxied us onto the runway, the engine humming eagerly beneath us.
>
> We began our ascent with a final radio call and a deep breath. As the plane picked up speed and the ground fell away beneath us, a rush of adrenaline surged through my veins. I watched in awe as the world below grew smaller and smaller, the horizon-expanding before me in a breathtaking panorama.
>
> The sensation was unlike anything I had ever experienced up in the air. The gentle hum of the engine, the steady thrum of the wind against the fuselage, and the feeling of weightlessness as we climbed higher. All combined to create a sense of freedom and exhilaration that words could scarcely capture.
>
> Under the guidance of my instructor, I took hold of the controls for the first time, feeling the responsiveness of the aircraft as I made slight adjustments to our course. Despite my initial nerves, I soon grew more confident with each passing moment, revelling in the sheer joy of flight.
>
> As we circled above, my instructor pointed out landmarks and taught me the basics of navigation, communication with air traffic control, and the importance of maintaining situational awareness. It filled me with a newfound sense of wonder and possibility.
>
> After what felt like an eternity and the blink of an eye, it was time to return to solid ground. With a series of gentle manoeuvres, we descended back toward the airstrip, the wheels touching down with a soft thud. As we taxied back to the hangar, I couldn't help but wear a grin that stretched from ear to ear. My first flying lesson was an experience I would never forget. As I stepped out of the cockpit, I knew this was just the beginning of an incredible journey into flight. There and then, I knew, this is what I was meant to do.
>
> *(Interview, 2023)*

The instructor's focus is on what and how the student is learning. In the first lessons, it makes the instructor's workload intense. The goal for the student is to be able to climb steadily while keeping the plane balanced and going in the right direction – knowing how to start and maintain a climb by using the right amount of engine power and controlling the plane's pitch, which is how high the nose is pointed. All the time, the instructor articulates, instructs, explains, supports, and encourages. An important learning point is how to level off the climb once the aircraft reaches a certain altitude. Together with the flight instructor, the student practices different types of climbs. One is climbing as fast as possible, another is climbing at a steeper angle, and the third is a climb that balances speed and altitude for efficient travel.

The student continued:

> in the following lessons we learn how to handle the plane when it's flying slower than usual. We do some checks before we slow down and then practice flying slowly while staying in control. We also learn how distractions can affect safety and how to get back to normal speed.

Even in the early exercises, students are being taught and tested, both to their own and to the plane's capacity:

> Stalls are when the plane's wings stop providing lift. We're taught how to recognize when a stall might happen and how to recover from it safely. We practice recovering from different types of stalls and learn how to react quickly if we get a stall warning.
>
> *(Interview, 2024)*

The second year builds on the fundamentals learned in the first, refining both the technical and cognitive aspects of piloting. Students deepen their understanding of aerodynamics, focusing on lift, drag, centre of gravity, and thrust. They explore stalls, slow flight, and aircraft handling characteristics, preparing for situations where control and recovery skills are required.

Meteorology now shifts from theoretical to operational decision-making. Students learn to analyse, recognizing how wind shear, pressure systems, and icing impact flight planning. They train in mass and balance calculations, ensuring that they can adjust for fuel loads, passenger distribution, and cargo placement to maintain aircraft stability.

Students are now introduced to operational procedures and emergency response training, where they master checklists, standard operating procedures, and troubleshooting techniques. These skills become second nature, ingrained through simulation and real-world application.

The summer phase marks another major transition as students begin instrument flight training (IR). Now, they learn to fly solely by reference to instruments, simulating low-visibility conditions, night flights, and instrument approaches. They develop absolute trust in their instruments, crucial for operating in Arctic conditions, where sudden weather shifts often obscure the horizon.

The third year marks a significant transition in pilot training, where students move from single-engine flight to multi-engine operations and complex flight environments. UiT has chosen to conduct the entire training (Instrument Flight Rules) on multi-engine aircraft, a decision influenced by the Arctic climate, where icing conditions may occur at any time of the year. In many training programmes, students initially complete their training on single-engine aircraft before transitioning to multi-engine flight. In the Arctic conditions, single-engine operations pose operational challenges, particularly in winter, when unpredictable icing and turbulent winds make de-icing capabilities and engine reliability a priority. UiT students complete approximately 50 hours of this training on multi-engine aircraft, significantly exceeding the minimum regulatory requirement of 16 hours.

This training is essential, as instrument flight proficiency is critical for Arctic aviation, where visibility can be severely reduced due polar weather and night conditions, and rapidly changing weather systems. Students develop confidence in their ability to fly solely by reference to instruments, navigating without external visual cues, and managing complex flight scenarios that require absolute trust in their instrumentation.

Simulator training plays a major role in this phase, exposing students to high-risk scenarios that cannot safely be replicated in live training flights. Students practice engine failures, hydraulic malfunctions, and forced landings, learning how to diagnose issues and execute procedures under pressure. This training is about developing the ability to remain composed and decisive in critical situations.

A major component of the final year of training is learning how to handle unexpected flight situations, including loss of control, through Upset Prevention and Recovery Training (UPRT). This training equips students with the skills needed to recognize, manage, and recover from extreme flight conditions. Although it represents only a limited portion of total flight training hours, its importance is widely recognized. The need for such training became evident following several high-profile accidents, most notably the crash of Air France Flight 447 in 2009 (BEA, Bureau of Enquiry and Analysis for Civil Aviation Safety, 2012), an event discussed at several points throughout this book. In that accident, the crew were unable to correctly interpret the flight conditions or recover from an aerodynamic stall. UPRT can be seen as a direct response to such failures, addressing gaps in understanding, perception, and manual handling under unexpected and rapidly evolving conditions. In a world where cockpit automation has an increasing role, pilots must still be able to manually recover from unexpected upsets. UiT students complete five hours of flight training along with five hours of theoretical instruction, ensuring they can recognize and respond to aircraft upsets before they escalate into critical situations.

The final stage of training in the UiT programme, like most other programmes, focuses on preparing students for the "reality" of commercial airline operations. While much of the early training emphasizes individual piloting skills, this introduces students to the complexities of coordinated cockpit decision-making. In commercial aviation, no flight is operated alone – pilots must work seamlessly as a team, communicate clearly, divide responsibilities between them, learn to lead and learn to follow, all while keeping a shared awareness. The term used in aviation for this type of training is Multi-Crew Cooperation (MCC). The whole idea behind it is to teach pilot students that flying is not a solo job. It is about working in a team

As students approach the end of their training, they complete their bachelor's thesis, selecting aviation-related topics that require critical analysis and problem-solving. This final academic requirement hopefully bridges some of the gap between theory and practice, allowing students to demonstrate their ability to engage with the broader aviation community and contribute to the evolving field of aviation and safety science.

9.5 PRACTICAL KNOWING AND LEARNING

In flying, learning, knowing, and doing are inextricably linked (Nergård, 1984). The knowledge involved is not merely transmitted through textbooks or theoretical instructions but embodied and experiential within flying as the ground zero of aviation. Pilot students are trained by partaking in a community of practitioners. The skills needed are honed through repeated action, observation, and reflection. The aim of education is to create all this and fit it together. The understanding of the different parts of the education finally happens in practical flying. A huge portion of it is tacit knowledge – unspoken, intuitive reflection. Learning to fly is not just about understanding aerodynamics or navigation charts, but a developing feeling for the aircraft, an intuitive sense of how it responds to different conditions, and the ability to make split-second decisions based on a deep, embodied understanding of the flying environment.

The process of learning what it takes to be part of this acting community begins with taking control of the aircraft and experiencing firsthand the sensations, challenges, and rewards of flying. The initial engagement is the foundation upon which further knowledge is built. Through continuous practice, pilots develop their skills, gradually internalizing the complex interplay of cognitive and physical tasks required to fly safely. This learning process is institutionalized through structured training programmes, mentorship, and a culture of continuous improvement. Knowledge is transferred from experienced pilots to novices through a combination of formal instruction and informal hands-on training. This transfer of knowledge is not a one-way street but a dynamic interaction where learners actively participate, ask questions, and engage in practice to refine their skills.

Learning is not a static attribute but a continuous development through engagement, reflection, and experience. Wittgenstein suggests that the verb "know" can be used analogously to the verbs "can" and "understand" (Wittgenstein 1978, § 150). This can be particularly illuminating for a pilot student. When we assert that someone "knows" how to fly an aircraft, we imply the competency that transcends the practical deed. In piloting, this means that knowledge encompasses both the intellectual grasp of aviation principles and the practical ability to perform them. Knowing how to execute a take-off involves more than memorizing the procedural checklist, as it includes performing the take-off smoothly and safely. Thus, "knows" implies a readiness to act, embodying both awareness and capability.

Similarly, when we say a pilot "understands" a particular flight technique, we are suggesting a depth of knowledge that allows flexible and adaptive application. Understanding signifies not just a comprehensive grasp of underlying principles but the practical execution of it at any moment. Understanding why specific manoeuvres are needed under particular weather conditions is crucial in the decision-making on a flight deck. It belongs to a pilot's overall proficiency and safety. As "understanding" must be transformed to "knowing," both indicate the demanded level of mastery on the flight deck.

During the education process in aviation, the students learn the language needed to deal with all practical operations. Learning the aviation language is important for the pilot student. It goes along with the mastery of flying an aircraft. Students learn it in their day-to-day training since there is no other way. Their language acquisition is learned as it is explained in their practical operations. The words and concepts in the aviation language appear on the exams that pilot students must pass to earn their certificates. These terms quickly become the first language through frequent use on every flight, while others may require memorization or reference to manuals and publications from governing entities that set the essential aviation terminology. Terms like "V1 speed," "rotate," or "holding pattern" have specific meanings on a flight deck and just there. The terms are embedded in the activities going on there. They are short hands for commands, orders, or acceptance. The pilot's understanding of them is deeply rooted in the reality of their operational environment. The language games of aviation are the specific contexts in which these terms are used, the protocols they trigger, and the actions they guide. The language and practices are inseparably connected, as the practice sets the meaning of it, and in turn asks for the relevant actions.

When outsiders witness the use of this language, it may appear poor until the working context of the shortcuts and the timeframe is fully understood, for example, when sitting in a cockpit and overhearing the shorthand dialogue between pilots.

In the cockpit context, the language used is precise and functional. A captain and a first officer share a common language game because they share a highly structured activity. Most joint actions consist of defined orders and corresponding responses, embedded in the complex operational universe of flying. In this setting, the meaning of words does not primarily lie in naming objects, but in directing action. Utterances function as instructions addressed to the other pilot. To say "descent checklist" is not to describe a checklist, but to initiate a sequence of coordinated actions:

> Take out the checklist for descent and carry out the points we must do in order to enter into the landing phase of this flight. Tell me in one word what you are doing at the same time as you are doing it.

Pilot students are trained in this knowledge and the corresponding skill. Knowing and doing is almost one and the same: "In order to master a landing, the student just achieve this skill through persistent practice and experience. That is why we use simulators in training; they are timesaving and a safer way of allowing students to train," an instructor told me (Interview, 2023). "You cannot read your way to a good landing. The ultimate demand for any pilot student is to transform their theoretical knowledge into practical proficiency" (Interview, 2023). For pilot students, achieving this mastery involves rigorous training, repeated practice, and a continuous learning process. When the pilot student becomes a licensed pilot, he or she is able to respond adeptly to real-world situations, but not until then. By then, they not only "know" the steps but are also able to perform them reliably.

The development of knowledge in aviation involves inexperienced pilots learning through trial and error in a pilot network that secures flight safety. The leading learning concept follows the basic formula: from grip to concept, in the words of Edvardsen (1996). The development of knowledge in the aviation tradition means skilled knowledge is always embodied in individual acting (Edvardsen, 2004: Merleau-Ponty 1994). Learning to become a competent pilot means being part of a skilful and highly competent culture in the same way as Wittgenstein argued that human activity belongs to a specific form of life (Wittgenstein, 1953, § 2, 23, 66).

The pilot student must become a part of the language and the activity of flight operations. The student must become an insider to the language games of which they are a part during each flight operation. During flight training, aspiring pilots undergo a socialization process where their theoretical and practical training form their perception and execution of flying. This experience is their cognitive framework supporting their practical skills when operating an aircraft. During their training, they add new experiences to the experiences that they already have. This forms a learning pattern the anthropologist Gregory Bateson (1972) named deutero-learning.

The concept is derived from the Greek "deuteros," meaning second or next. Bateson distinguishes two levels of learning: proto-learning and deutero-learning, but both occur simultaneously. Proto-learning involves acquiring specific skills or knowledge, such as riding a bike, learning a language, or repairing a car. Deutero-learning relates to the broader learning process where the proto-learning melt together with the learning process of the second order (Visser, 2003).

For pilot students, deutero-learning forms and hosts the reference of the experienced piloting in which their theoretical and practical training advance to a new level of meaning, influencing not just their immediate learning outcomes, but also their overall knowledge of the complexities of aviation. It is the combination of the two levels of learning that allows the student to make progress in the learning process. Theoretical and practical training becomes a joint venture when proto-learning advances into deutero-learning.

To become a pilot involves the preparation for a certain state of being. I many respects, it involves a personal journey grounded in the essence of flying. The acquisition of practical flying skills is inseparably tied to adopting a specific mindset deeply ingrained in the pilot community. What is additionally needed to extend all complicated practical knowledge is the human sense needed to master its lack of will and intentions. "You must become the size of your aircraft," a captain said. "It's like putting on your pants. When you put on your seat belts, you put on the aircraft like you do with pants. Your body must become the size of the aircraft" (interview, 2004).

The essence of knowing how to fly an aircraft is the performative act of flying. The knowledge needed is embodied in the fusion of practical skill and creative know-how.

It is not merely a collection of will and facts but is deeply rooted in the mastery of a specific form of life in the aviation industry. It involves a specific set of skills, practices, and situational understandings. To truly know how to fly an aircraft is to perform a practice, make judgments from within, and respond to the ever-changing conditions of a flight. Flying an aircraft is not just about following procedures or understanding aerodynamic principles. It involves a profound, embodied understanding that comes from experience. This practical knowledge is a form of art. It must, at the same time, consist of knowing what to do in specific situations, often without being able to fully articulate the reasoning behind each part of the acting, at least not in the very moment it takes place (Molander, 1996). In some moments, it appears as an intuitive grasp of the dynamics of flight, weather conditions, the aircraft's behaviour, and the myriad of variables that may affect an actual flight.

In aviation, therefore, knowledge must be dynamic, not abstract but concrete. It must be a sort of wisdom gained from experience and practice, evolving creative know-how that enables a pilot to adapt to unforeseen challenges and situations. This knowledge needs to be deeply contextual and situational, relying on the pilot's ability to integrate sensory information, technical data, and past experiences into coherent and effective actions.

Flying an aircraft is thus as intellectual as practical, blending theoretical understanding with practical engagement. It is the embodiment of Wittgenstein's idea that the limits of our language are the limits of our world, and Molander's idea of the profound, silent knowledge that guides the acting. The essence of knowing how to fly is found in the seamless integration of thought and action, theory and practice, knowledge and wise judgment.

This knowledge exists within the aviation community and becomes apparent through actual practice (Meløe, 1979). This type of knowledge is shared and transferred among insiders and often articulated among them in a language code that to the outsider may appear as poor language. Among the community of insiders, it works perfectly well as they grasp it as a "complete language," as Wittgenstein puts it. To fly an aircraft from one destination to another is the essential form of aviation practice. The pilot's part of this practice is to fly the aircraft within the operational and regulatory frameworks established by the aviation industry.

9.6 FLYING IN THE ARCTIC

The Arctic is not just a location for pilot training. It is an ever-present teacher, demanding constant attentiveness and adaptability. Learning to fly in this region means learning to operate in an environment that cannot be fully foreseen and controlled. Rapidly shifting weather patterns, rugged terrain, and extreme cold require pilots to read the skies, anticipate turbulence, and understand how temperature, wind, and ice affect aircraft performance.

The Arctic brings together different weather systems. Warm, humid air from the Gulf Stream collides with cold polar winds, creating unpredictable conditions. Ice on an aircraft affects lift, control surfaces, and engine performance. This is not a slow process and may develop within minutes. Pilot students learn to recognize the first signs of icing and react before the instruments confirm the problem. This requires technical knowledge and demands intuition, as well as the ability to make quick decisions in dynamic situations.

Cold temperatures affect aircraft in other ways. Hydraulic systems respond differently, fuel thickens, and metal contracts, changing the aircraft's handling. What would be a minor adjustment in warmer conditions becomes crucial in freezing temperatures. Students training for the north learn to sense these changes and handle the aircraft's behaviour as part of a dynamic interaction between machine and environment, as seen in Figure 9.2.

Flying in such conditions is not just an intellectual exercise. It is a hands-on experience where the aircraft becomes an extension of the pilot's awareness. The plane is not merely a machine to be operated mechanically but also one to be listened to with sense before conscious thought. The aircraft speaks its own language – a change in engine tone, a shift in air pressure, or resistance in the control surfaces. These are signals experienced pilots sense instinctively long before instruments confirm them.

This connection between pilot and aircraft is not about control. It is about understanding and adapting. It is an ongoing process where pilots develop a feel for the machine – its rhythms, signals, and responses. This is not a skill set that can be learned from manuals alone. It is developed through structured training, repetition, and real-world exposure. More than technical competence, it becomes a way of thinking, where precision, perception, and adaptability merge into an operational mindset.

An instructor explained:

> We teach in an environment where the weather can change in minutes. The students have to learn to trust their instruments because visual references can disappear in a whiteout. Taxiing on icy runways is a challenge many don't expect, but mastering it builds confidence.
>
> *(Interview, 2024)*

FIGURE 9.2 Arctic weather.

(Photo: Christian Aasen Stratton.)

These experiences develop students not only as pilots but as decision-makers who learn to interpret complex situations and act with confidence. The Arctic teaches them special lessons to be aware of the unexpected and to adjust to it without hesitation.

The Arctic is more than just a harsh environment. It is a site where theory and practice are asked to come together in a way few other settings offer. In the lecture hall, students learn about wind patterns, air pressure, and the principles of lift and drag. In the cockpit, they feel these forces firsthand. What begins as an equation on a whiteboard becomes a lived experience as the aircraft responds to gusts, turbulence, and shifting air currents.

This constant movement between theory and practice is not without tension. In the classroom, students are encouraged to question, debate, and analyse. In the cockpit, there is less room for doubt. Decisions must be made quickly and executed precisely. When a crosswind pushes the aircraft off course, a student cannot pause to reflect. They must correct immediately, instinctively, and with absolute confidence.

This creates a unique way of thinking that balances reflection and action. Students learn to analyse situations deeply, but they also learn to act decisively when needed. They learn to carry knowledge into practice and to learn from each flight experience.

There is no way to fully simulate Arctic flying in a classroom. Students must feel the plane struggling against strong winds and navigate icy runways with reduced friction. These challenges are not obstacles to learn as they are the learning itself.

Through repetition and experience, students gain the confidence to manage turbulence, land in crosswinds, and know when to abandon an approach.

This is where the Arctic plays its most significant role. It does not allow students to rely solely on procedures or checklists. It asks them to develop judgment, adaptability, and resilience. In this environment, students learn to read, interpret, and respond to a world that refuses to be simplified. They learn to operate not just as pilots but as aviators who can navigate the unpredictable.

The UiT programme brings something new to the table. Theory and practice are not separate stages. They are woven together in a continuous cycle. When students move between the lecture hall and the cockpit, they carry knowledge into real-world scenarios, testing and reshaping it through experience. The Arctic makes this learning procedure unique. Here, theory is not just learned; it is experienced, felt, lived, and applied under unique circumstances.

This is the essence of pilot education at UiT. In the Arctic, this cycle of learning is not merely instructional – it is a necessity. It is within this demanding environment that local and universal knowledge is transformed into embodied skills, a way of thinking and acting that defines the profession in the Arctic landscape (Figure 9.3).

FIGURE 9.3 Final approach. The quiet moment where training, judgement, and trust in each other merge into a single movement.

NOTE

1 The pilot education programme at UiT The Arctic University of Norway is fully aligned with international aviation standards and Norwegian higher education regulations. It adheres to EASA's Aircrew Regulation (EU) No. 1178/2011, ensuring compliance with European pilot licensing requirements, and follows the Bologna process, securing degree recognition across the European Higher Education Area. Nationally, it is accredited by NOKUT and meets the Norwegian Qualifications Framework for Lifelong Learning (NQF), ensuring that academic and practical training adhere to both aviation and higher education quality standards. This alignment guarantees that UiT graduates are professionally qualified and internationally recognized.

10 Man and Machine
Concluding Remarks on the Ethnography of Practice

As soon as the aircraft has left the ground, it is in the hands of the pilots and crew. The other perspective is also the case: the aircraft is in the hands of the crew when it has left the ground. When airborne, the two share a destiny. Especially for the pilots, the flight means intense interaction between man and machine. They act on each other and demand certain reactions from each other, but only the pilot understands in a human sense what is going on and what is at stake. The machine continuously reports whether it works normally or the opposite if it does not.

A kind of dialogue between man and machine is going on during a flight. In few other work occupations do humans and machines so basically share a common destiny. Pilots may not grasp every microscopic detail of the aircraft's systems, but they must develop a working comprehension of how its essential components function, how they interact, and how they might fail.

I have already introduced the idea that the machinery during flight operations appears to pilots as an extension of their own mind and body. Pilots act on the machine, but they also act through it. The will of the pilot somehow becomes the will of the aircraft. This nearness and perception do not disturb the pilot's advanced professional understanding of the technical order of the aircraft. It rather brings a human element into it and to the relation and handling of the machine. It may be considered as a way of being familiar with the advanced machinery and coming to terms with it. But it is also rooted in a relation over which humans have limited control when it comes to advanced technology.

The tendency to humanize the relation between man and material can be recognized in a lot of different practices. It is the human way of coming to terms with complicated technology or nature that we do not fully understand. It is bound to the relation itself. The handling of advanced technology, which humans are supposed to control, may motivate the establishment of imaginary dialogues with technical challenges or with nature. It is not a superstitious way of approaching complex elements in human and nonhuman environments, but rather a way to bring the human operator close to the challenge involved. The humanizing mode of handling the relationship between man and machine may improve human perception of the task and the mastery of it. It contributes to a mode of adequate operational comprehension with practice as a plausible way of handling work operations without knowing the complete order of it.

In cultural studies of life and living closely bound to nature, we recognize this phenomenon. In traditional nature-based forms of living, knowledge and experience are gained from operating in nature when harvesting it. The community dependent

DOI: 10.1201/9781003725862-10

on a certain nature must know it well and know the limits of harvesting it in order to make nature able to reproduce itself. The practices and their use therefore ask for a deep understanding of the limits of nature. The community living from its resources has, over generations, gained and developed the knowledge needed to survive and have a life there. People living there know well that the nature they employ has its own order and limits, and they also know they need to fully respect this order. They share destiny with the nature they harvest and have at their disposal, but which they still do not fully understand. If these resources decline, the harvesting will gradually ruin the very foundations of the community itself.

The harvesting community knows much about the order of nature they use and employ. Still, these limits cannot be fully known since the order of nature is continuously shifting. This is the very nature of the relationship between man and nature. Nature changes both for internal reasons and also because of human invention in its conditions. The continuously shifting elements of nature hence cannot be fully known to humans in advance, no matter how experienced the community of harvesting is. Just because the community knows nature well, it also knows that every aspect of nature's order cannot be foreseen and understood in advance. It also knows that the human use of it alters its character.

Therefore, nature always leaves some aspect of its order unknown to humans, and these are yet to be discovered. There is always something unknown about nature hidden from the harvesting community. Nature therefore continuously leaves the community with wonder about the way it works, under its own changing circumstances, partly coming from the way humans intervene in nature in order to improve their living conditions. Nature changes community life continuously by the way it sets new conditions for the harvesting practices. Community life must approach nature in new ways. This shift in harvesting and use of nature reveals just a new side of nature's own complexity. In order to deal with this change, the community must develop a system in which nature has a language of its own based on human experience with nature (Nergård, 2019). The cultural storage of this experience makes nature able to speak up for itself and avoid the human community from making harm to it (Nergård, 2021).

The Sámi tradition has developed a system of dealing with nature in which nature speaks up for itself. For a long period of history, this tradition was seen as superstitious and not fully developed culturally. The traditional Sámi approach to nature was indeed not a "primitive" or irrational idea, but quite on the contrary, an experience-based approach to nature. The consequences of misconduct in Sámi nature were turned into norms for acting in nature meant to prevent them. It was a system of knowledge in which nature spoke up for people on a spot of common interests, in order to remind them about the limits of nature. In Sámi tradition, any particular river or lake rich in trout, an area rich in cloudberry, or a site known to be a good hunting ground for grouse, has narratives reminding people about being cautious with nature at such sites. Even if modern science may explain the balance of ecology of such places, the system still works alongside modern concepts of ecology and nature conservation in order to care for nature.

The traditional experience-based culture has an internal order whose knowledge cannot be questioned from the outside. It is a certain form of life existing in its own right as any other. From the outside that it may appear as lacking something or being unintelligible. We therefore must distinguish between how such forms of life or practices appear from the outside and how they appear to the community of insiders. This holds whether we talk about rural nature-based communities or advanced technical communities such as those we find in the aviation industry, or at a flight deck, for that matter. Their goals and destiny may be different, but they are specimens of the same logical order when it comes to their internal life. That holds for their language and coding of action as part of what is going on within their everyday life.

I have introduced my reader to the cockpit communication and the shorthand language used on the flight deck. I have underlined that the shorthand language belongs to highly skilled and professional acting. It is contrary to the manifestation of poor language and limited understanding. It works and confirms the adequate presence of skilled acting.

It is the language of doing more than that of explaining. Most of the time it works unlike the language of the university seminar, in which explaining something is the goal of the activity itself. On the flight deck, the language just accompanies the coordination of work on advanced technology. The goal of the language is not explaining, but, quite on the contrary, to support procedural acting. As in other practices, this language takes and gets its meaning from the activity to which it belongs. We may say the rationality of this language is part of the rationality of acting in the practice to which it belongs.

Jakob Meløe (1988) argues:

> The method of investigating *the concept* of a harbour, therefore, is this: Situate yourself within the practice that this *object* belongs to and then investigate *the object* and *its* contribution to that *practice*. If an object belongs essential to a practice, as a harbour does, (...), then the concept of that object is our understanding of that object's contribution to the practice within which it is that object (...).

The aircraft and its systems may fail. In aviation history, many of the accidents that we have experienced, especially in the early days of aviation, can be attributed to failures of the machine or its systems. During the last decades or so, for the most part of it is the human who is the cause of failure with tragic outcomes. At times and in some cases, it is the interaction between the human and the machine that seems to cause most of the accidents. The balance between advanced technology and the human ability to cope with it will be a huge challenge for research and training in future aviation education.

Any profession is more than a collection of skills. It is a way of seeing the world and acting within it. It is an understanding that grows through practice until it becomes part of how one moves and thinks. Reindeer herders follow their animals and read the land through them. What they know is shown in the way they act. Surgeons do more than perform procedures. Their hands learn the small differences in tissue and pressure and the rhythm that guides each step. Pilots also move beyond technical knowledge. They develop a feel for their aircraft and their crew and for the air they work in. In all these practices the meaning lies in the doing. Knowledge is not something separate from the work. It is lived. Airmanship belongs here. The craft rests on the quiet judgement of those who shoulder the responsibility and act when the moment calls for it.

Bibliography

Alkov, R. A., Borowsky, M. S., Williamson, D. W., & Yacavone, D. W. (1992). The effect of trans-cockpit authority gradient on Navy/Marine helicopter mishaps. *Aviation, Space, and Environmental Medicine*, *63*(8), 659–661.

Anderson, D. (2012). Is building relationships the key to leadership? *Performance Improvement*, *51*, 15–22.

Ash, J. (2023). *Helicopter operations in Svalbard* [Personal communication].

Aviation Safety Network. (2007). *CVR transcript Avianca Flight 052 - 25 JAN 1990*. Retrieved 25 January 2025, from https://web.archive.org/web/20210730033840/

Bainbridge, Lisanne (1983). Ironies of automation. *Automatica*, *19*(6), 775–779.

Baker. (2002). *Se s.* 128.

Balcerzak, T., Kostur, K., & Rajchel, J. (2024). The "Black Swan" theory in aviation safety. *Ius Novum*, *18*(3), 34–46. https://doi.org/10.2478/in-2024-0021

Bartone, P. T. (2006). Resilience under military operational stress: Can leaders influence hardiness? *Military Psychology*, *18*(S), S131–S148. https://doi.org/10.1207/s15327876mp1803s_10

Bateson, G. (1972). *Steps to an ecology of mind*. New York: Ballantine Books.

Batteau, A. W. (2001). The anthropology of aviation and flight safety. *Human Organization*, *60*(3), 201–211. https://doi.org/10.17730/humo.60.3.3q5m5dylr01p1qa4

Benjamin, W. (1991). *Kunstverket i reproduksjonsalderen. Essay om kultur, litteratur og politikk*. Oslo: Gyldendal Norsk Forlag.

Bergström, J., Dahlström, N., Dekker, S., & Petersen, K. (2011). Complicated and complex systems: What would successful safety interventions look like? In J. Pariès & J. Wreathall (Eds.), *Resilience Engineering in Practice: A Guidebook* (1st ed.). London: CRC Press. https://doi.org/10.1201/9781317065265

Bernstein, B. (1996a). *Pedagogy, symbolic control, and identity: Theory, research, critique*. Bristol, PA: Taylor and Francis.

Bernstein, P. L. (1996b). *Against the gods: The remarkable story of risk*. New York: John Wiley & Sons.

Bion, W. R. (1962). *Learning from experience*. London: Heinemann.

Bion, W. R. (1963). *Elements of psycho-analysis*. Heinnemann.

Bion, W. R. (1970). *Attention and interpretation: A scientific approach to insight in psycho-analysis and groups* (p. 136). London: Tavistock.

Bion, W. R. (2018). *Transformations* (1st ed.). Routledge.

Bor, R., Eriksen, C., Hubbard, T., & King, R. (Eds.). (2019). *Pilot selection: Psychological principles and practice* (1st ed.). CRC Press.

Boston College. (n.d.). *The evolution of aviation*. Retrieved 25 August 2025, from https://bostoncolleges.in/the-evolution-of-aviation/

Brady, T. (2000). Pilot education: The beginnings. *Journal of Aviation/Aerospace Education & Research*, *9*(2). https://doi.org/10.15394/jaaer.2000.1251

Braithwaite, G. (2001a). *Attitudes and reality: An investigation into the attitudes and practices of aircraft maintenance personnel and the implications for safety*. Aldershot: Ashgate.

Braithwaite, G. R. (2001b). *Attitude or latitude? Australian aviation safety*. Aldershot: Ashgate Publishing, United Kingdom.

Bronfenbrenner, U. (1996). *The ecology of human development: Experiments by nature and design* (1st ed.). Harvard University Press.

Bureau d'Enquêtes et d'Analyses pour la sécurité de l'aviation civile (BEA). (2012). *Final report on the accident on 1 June 2009 to the Airbus A330-203 registered F-GZCP operated by Air France flight AF 447 Rio de Janeiro – Paris*. Paris: BEA. https://www.bea.aero

Cameroon Civil Aviation Authority (CCAA). (2010). *Technical investigation into the accident of the B737-800 registration 5Y-KYA operated by Kenya Airways that occurred on 5 May 2007 in Douala*. Retrieved 27 August 2025, from https://asn.flightsafety.org/reports/2007/20070505_B738_5Y-KYA.pdf

Cannon-Bowers, J. A., Salas, E., & Converse, S. A. (1990). Cognitive psychology and team training: Shared mental models in complex systems. *Human Factors Bulletin*, *33*, 1–4.

Chatzi, A. V., Martin, W., Bates, P., & Murray, P. (2019). The unexplored link between communication and trust in aviation maintenance practice. *Aerospace*, *6*(6), 66. MDPI AG. http://dx.doi.org/10.3390/aerospace6060066

Clay-Williams, R., Austin, E., Braithwaite, J., Hollnagel, E., & Rapport, F. (2021). Qualitative assessment to improve everyday activities: Work-as-imagined and work-as-done. In *Transforming healthcare with qualitative research* (1st ed., pp. 71–82). Routledge. https://doi.org/10.4324/9780429299797-11.

Cookson, S. (2019). Intercultural communication on the flight deck: A review of studies in aviation. In N. Stanton (Ed.), *Advances in human aspects of transportation (AHFE 2018). Advances in Intelligent Systems and Computing, 786*. Cham: Springer. https://doi.org/10.1007/978-3-319-93885-1_8

Dahlström, N., Dekker, S., van Winsen, R., & Nyce, J. (2009). Fidelity and validity of simulator training. *Theoretical Issues in Ergonomics Science*, *10*(4), 305–314.

Damos, D. L., John, R. S., & Lyall, B. (1999). Pilot error: Implications of automation and workload. In R. Jensen & L. Rakovan (Eds.), *Proceedings of the 10th International Symposium on Aviation Psychology* (pp. 326–332). Columbus, OH: The Ohio State University.

Dekker, S. (2014). *The field guide to understanding human error*. London: CRC Press.

Dismukes, R. K. (Ed.). (2009). *Human error in aviation*. London: Routledge.

Dismukes, R. K., & Smith, G. M. (2000). *Facilitation and debriefing in aviation training and operations*. New York: Routledge.

Dismukes, R. K., Berman, B. A., & Loukopoulos, L. D. (2007). *The limits of expertise: Rethinking pilot error and the causes of airline accidents*. Aldershot: Ashgate Publishing Limited.

Dupont, G. (1997). The dirty dozen errors in aviation maintenance. In *Proceedings of the 11th federal aviation administration meeting on human factors issues in aircraft maintenance and inspection: Human error in aviation maintenance* (pp. 45–49). Washington, DC: Federal Aviation Administration/Office of Aviation Medicine.

EASA. (n.d.). https://www.easa.europa.eu/

Edvardsen, E. (1996). *Den gjenstridige allmue: Skole og levebrød i et nordnorsk kystsamfunn ca 1850–1900* (p. 160). Oslo: Solum.

Edvardsen, E. (2004). *Samfunnsaktiv skole – En skole rik på handling*. Lillehammer: Oplandske Bokforlag.

Endsley, M. R. (2015). Situation awareness misconceptions and misunderstandings. *Journal of Cognitive Engineering and Decision Making*, *9*(1), 4–32. https://doi.org/10.1177/1555343415572631

European Parliament and Council. Regulation (EC) No 1008/2008 of 24 September 2008 on common rules for the operation of air services in the Community. Official Journal of the European Union, L 293/3–20. http://data.europa.eu/eli/reg/2008/1008/2020-12-18

Everly, G. S., Flannery, R. B., & Mitchell, J. T. (2000). Critical incident stress management (CISM). *Aggression and Violent Behavior*, *5*(1), 23–40. https://doi.org/10.1016/S1359-1789(98)00026-3

Ferris, T., Sarter, N., & Wickens, C. D. (2007). Cockpit automation: Still struggling to catch up. In R. Parasuraman & M. Rizzo (Eds.), *Neuroergonomics: The brain at work* (pp. 391–406). Oxford: Oxford University Press.

Geertz, C. (1974). From the native's point of view: On the nature of anthropical understanding. *Bulletin of the American Academy of Arts and Sciences, 28*(1), 26–45.

Geertz, C. (1983). *Local knowledge: Further essays in interpretive anthropology*. New York: Basic Books Inc.

Geertz, C. (1984). Anti anti-relativism. *American Anthropologist, 86*(2), 263–278.

Georgiou, A. M., Littlepage, G. E., & Hein, M. B. (2021). What we know about teamwork and multiteam coordination in aviation: Contextual factors affecting teamwork in aviation. *CORE Scholar International Symposium on Aviation Psychology*, Dayton, OH.

Goffman, E. (1974). *Vårt rollespill til daglig: En studie i hverdagslivets dramatikk*. Oslo: Dreyer Forlag.

Greenberg, J. R., & Mitchell, S. A. (1983). *Object relations in psychoanalytic theory*. Harvard University Press. https://doi.org/10.2307/j.ctvjk2xv6

Hall, E. T. (1963). A system for the notation of proxemic behavior. *American Anthropologist, 65*(5), 1003–1026. http://www.jstor.org/stable/668580

Heinrich, H. W. (1931). *Industrial accident prevention: A scientific approach*. New York: McGraw-Hill.

Helmreich, R. L., & Foushee, H. C. (1993). Why crew resource management? Empirical and theoretical bases of human factors training in aviation. In E. L. Wiener, B. G. Kanki, & R. L. Helmreich (Eds.), *Cockpit resource management* (pp. 3–45). San Diego, CA: Academic Press.

Helmreich, R. L., & Merritt, A. C. (2001). *Culture at work in aviation and medicine: National, organizational, and professional influences*. London: Routledge.

Helmreich, R. L. (1998). *Managing human error in aviation*. Aldershot: Ashgate.

Helmreich, R. L., Merritt, A. C., & Wilhelm, J. A. (1999). The evolution of crew resource management training in commercial aviation. *The International Journal of Aviation Psychology, 9*(1), 19–32. https://doi.org/10.1207/s15327108ijap0901_2

Hollnagel, E. (2010). *The ETTO principle: Efficiency thoroughness trade-off*. Farnham: Ashgate.

Hollnagel, E. (2013). A tale of two safeties. *Nuclear Safety and Simulation, 4*(1), 1–9.

Hollnagel, E. (2014). *Safety-I and Safety-II: The past and future of safety management*. Boca Raton, FL: Taylor & Francis Ltd.

Hollnagel, E., Pariès, J., Woods, D. D., & Wreathall, J. (2010). *Resilience engineering in practice: A guidebook* (pp. 243–257). Ashgate.

Hutchins, E., Holder, B., & Perez, R. (1999). The pilot–aircraft–system team: Overcoming the ironies of automation. In D. J. Garland, J. A. Wise, & V. D. Hopkin (Eds.), *Handbook of aviation human factors* (pp. 3–30). Mahwah, NJ: Lawrence Erlbaum Associates.

International Civil Aviation Organization. (2006). *Safety management manual (SMM) (Doc 9859)*. Montreal: ICAO.

Jung, C. G. (2007). *The Archetypes and the Collective Unconscious*. New York: Routledge.

Kanki, B. G., & Palmer, M. T. (1993). Communication and crew resource management. In E. L. Wiener, B. G. Kanki, & R. L. Helmreich (Eds.), *Cockpit resource management* (pp. 99–136). San Diego, CA: Academic Press.

Karlsson, B., & Borg, M. (2013). *Psykisk helsearbeid: humane og sosiale perspektiver og praksiser*. Oslo: Gyldendal Akademisk.

Kern, T. (1997). *Redefining airmanship*. New York: McGraw Hill.

Klein, M. (1957). *Envy and gratitude: A study of unconscious forces*. London: Hogarth Press.

Klein, M. (1988). *Kärlek, skuld och gottgörelse*. Stockholm: Natur och Kultur.

Kolluru, R., Bartell, S. M., Pitblado, R. M., & Stricoff, R. S. (1996). *Risk assessment and management handbook for environmental, health, and safety professionals*. New York: McGraw Hill.

Lamplugh, A. G. (c. 1930s). *Quotation: Flying is not inherently dangerous....* Widely attributed to Captain Alfred Gilmer Lamplugh, British Aviation Insurance Group. [No formal publication available].

Leach, E. (1982). *Social anthropology*. Glasgow, UK: Fontana Paperbacks.

Lear, J. (2009). Technique and final cause in psychoanalysis: Four ways of looking at one moment. *The International Journal of Psychoanalysis*, *90*, 1299–1317. https://doi.org/10.1111/j.1745-8315.2009.00204.x

Lindbergh, Charles A. (1927). *We*. New York: G. P. Putnam's Sons.

Malcolm, A. S., Seaton, J., Perera, A., Sheehan, D. C., & Van Hasselt, V. B. (2005). Critical incident stress debriefing and law enforcement: An evaluative review. *Brief Treatment and Crisis Intervention*, *5*(3), 261–278. https://doi.org/10.1093/brief-treatment/mhi019

Mallinson, H. (2018). *Flight secrets: Flight attendant reveals crew had to look like THIS during flying heyday*. Retrieved 18 August 2025, from https://www.express.co.uk/travel/articles/996495/flight-secrets-cabin-crew-air-hostess-1960s-plane

Martinussen, M. (1996). Psychological measures as predictors of pilot performance: A meta-analysis. *The International Journal of Aviation Psychology*, *6*(1), 1–20. https://doi.org/10.1207/s15327108ijap0601_1

Martinussen, M. (2005). Seleksjon av flygere og flygeledere. *Tidsskrift for Norsk psykologforening*, *42*(4), 291–300. https://www.psykologtidsskriftet.no/artikkel/2005as03ae-Seleksjon-av-flygere-og-flygeledere

Martinussen, M., & Hunter, D. R. (2017). *Aviation psychology and human factors*. Boca Raton, FL: Taylor & Francis.

McFee, G. (2012). *Sport, rules and values: Philosophical investigations into the nature of sport*. London: Taylor & Francis.

Meløe, J. (1974). «Aktøren og hans verden.» *Norsk filosofisk tidsskrift*, *2/1973*, 133–143.

Meløe, J. (1979). *Notater i vitenskapsteori for gruppene humaniora og samfunnsvitenskap med fiskerifag*. Tromsø: Universitetet i Tromsø.

Meløe, J. (1988). «The two landscapes of Northern Norway.» *Inquiry*, *3/1988*, 387–401.

Meløe, J. (1997). Om å forstå det andre gjør. In *Filosofi i et nordlig landskap: Jakob Meløe 70 år*. Tromsø: Universitetsbibliotek i Tromsø.

Meløe, J. (n.d.). *Steder*. Retrieved 16 August 2025, from https://www.jakobmeloe.com/files/steder3.pdf

Merleau-Ponty, M. (1994). *Kroppens fenomenologi* (pp. XII, 195). Oslo: Pax Forlag.

Merritt, A. (1993a). Cultural values in pilot judgment and decision making. In R. S. Jensen & D. Neumeister (Eds.), *Proceedings of the Seventh International Symposium on Aviation Psychology* (pp. 682–687). Columbus, OH: Ohio State University.

Merritt, A. C. (1993b). *The influence of national and organizational culture on human performance. Australian Aviation Psychology Association Seminar*, Sydney, Oct 25, 2003.

Merritt, A., & Helmreich, R. L. (1996). Human factors on the flight deck: The influence of national culture. *Journal of Cross-Cultural Psychology*, *27*(1), 5–24.

Moira, P., Mylonopoulos, D., & Terzoglou, E. (2020). A typology of delinquent behavior on the aircraft. *Journal of Tourism and Leisure Studies*, *5*(1), 29.

Molander, B. (1996). *Kunnskap i handling* (2. omarbetade uppl., p. 286). Daidalos.

Moloney, K. (n.d.). *Elevate Her Aviation: Do words matter? Cockpit vs flight deck*. Retrieved 25 August 2025, from https://elevateheraviation.co.uk/about-the-founder/f/do-words-mattercockpit-vs-flight-deck

NASA. (1979). H. P. Ruffell Smith. *A simulator study of the interaction of pilot workload with errors, vigilance, and decisions*. NASA Technical Memorandum 78482. Ames Research Center, Moffett Field, California.

National Transportation Safety Board. (1991). *Aircraft accident report, Avianca, the Airline of Colombia, Boeing 707-321B, HK 2016, fuel exhaustion, Cove Neck, New York, January 25, 1990. (NTSB/AAR-91/04)*. Washington, DC: National Transportation Safety Board. Retrieved 15 January 2025, from https://www.ntsb.gov/investigations/accidentreports/reports/aar9104.pdf

National Transportation Safety Board. (2007). *Attempted takeoff from wrong runway, Comair Flight 5191, Bombardier CL-600-2B19, N431CA, Lexington, Kentucky, August 27, 2006*. (Aircraft Accident Report NTSB/AAR-07/05). Washington, DC. Retrieved 27 August 2025, from https://www.ntsb.gov/investigations/AccidentReports/Reports/AAR0705.pdf

Navarro, J. (2008). *What everybody is saying*. New York: Harper Collins Books.

Nergård, J.-I. (1984). *Språk og praksisformer: om språkets plass i handling, forståelse og sosial forandring*. Tromsø: UIT Norges arktiske universitet, Universitetsbiblioteket.

Nergård, V. (2005). *Kinship and ritual kinship in Sámi communities – Contributions to a psychodynamic understanding of socialisation* (Doctoral dissertation). University of Oslo.

Nergård, V. (2014). Airmanship – a qualitative approach. *Aviation, 18*(3), 147–156. https://doi.org/10.3846/16487788.2014.969882

Nergård, V. (2015). Learning from experience. *Aviation (Vilnius, Lithuania), 19*(4), 171–179. https://doi.org/10.3846/16487788.2015.1127452

Nergård, J.-I. (2019). *Dialoger med naturen: etnografiske skisser fra Sápmi*. Oslo: Universitetsforlaget.

Nergård, J.-I. (2021). *The Sámi narrative tradition*. New York: Taylor & Francis.

Nergård, V., & Ash, J. (2020). Selection of air ambulance pilots: Empathic airmanship and the safe pilot. In *Pilot selection* (pp. 331–339). New York: CRC Press.

Nergård, V., Hatlevik, O. E., Martinussen, M., & Lervåg, A. (2011). An airman's personal attitude: Pilots' point of view. *Aviation, 15*(4), 101–111. https://doi.org/10.3846/16487788.2011.651789

Orasanu, J. M. (1994). "Shared problem models and flight crew performance." In N. Johnston, N. McDonald, & R. Fuller (Eds.), *Aviation psychology in practice*. AldershotAvebury Technical.

Parasuraman, R., & Riley, V. (1997). Humans and automation: Use, misuse, disuse, abuse. *Human Factors, 39*(2), 230–253. https://doi.org/10.1518/001872097778543886

Parret, H. (1994). Indirection, manipulation and seduction in discourse. In Herman Parret (Ed.), *Pretending to communicate*. Berlin, Germany: Walter De Gruyter GmbH.

Paul, H. (2011). Distance and self-distanciation: Intellectual virtue and historical method around 1900. *History and Theory, 50*, 104–116. https://doi.org/10.1111/j.1468-2303.2011.00606.x

Perrow, C. (1984). *Normal accidents: Living with high-risk technologies*. New York: Basic Books.

Petrescu, R. V. V., Aversa, R., Akash, R., Bucinell, R., Apicella, A., & Petrescu, F. I. T. (2017). *History of aviation: A short review. Journal of Aircraft and Spacecraft Technology, 1*(1), 30–49. https://doi.org/10.3844/jastsp.2017.30.49

Poirot, L. (2024). Airplane cabins through the years. Retrieved 25 January 2025 from https://www.farandwide.com/s/airplane-cabins-history-timeline-011caad44b6e4eb9.

Popham, A. R. (1957). *The drawings of Leonardo da Vinci*. London: The Reprint Society.

Reason, J. (1997). *Managing the risks of organisational accidents*. Aldershot: Ashgate.

Roschelle, J., & Teasley, S. D. (1995). The construction of shared knowledge in collaborative problem solving. In C. O'Malley (Ed.), *Computer supported collaborative learning (NATO ASI Series)*, Vol. 128, pp. 69–97). Berlin, Heidelberg: Springer. https://doi.org/10.1007/978-3-642-85098-1_5

Ruesch, J., & Bateson, G. (1968). *Communication: The social matrix of psychiatry*. New York: Norton.

Sarter, N. B., & Woods, D. D. (1997). Team play with a powerful and independent agent: Operational experiences and automation surprises on the Airbus A-320. *Human Factors, 39*(4), 553–569. https://doi.org/10.1518/001872097778543912

Schröder-Hinrichs, J. U., Hollnagel, E., & Baldauf, M. (2012). From *Titanic* to *Costa Concordia*: A century of lessons not learned. *WMU Journal of Maritime Affairs, 11*, 151–167.

Segal, H. (1989). *Klein* (p. 189). London: Karnac Books and the Institute of Psychoanalysis.
Sennett, R. (2008). *The craftsman*. London: Penguin Books.
Shawver, L. (1996). What postmodernism can do for psychoanalysis: A guide to the postmodern vision. *The American Journal of Psychoanalysis, 56*(4), 371–394. https://doi.org/10.1007/BF02735489
Smith, H. P. R. (1979). *A simulator study of the interaction of pilot workload with errors, vigilance, and decisions*. NASA Technical Memorandum 78482. California: Ames Research Center, Moffett Field, California.
Stierlin, H. (1974). *Konflikt og forsoning: Om dynamikken i mellommenneskelige forhold*. Oslo: Gyldendal Norsk Forlag; Lund University Press (1991).
Strong, T., Sutherland, O., & Ness, O. (2011). Considerations for a discourse of collaboration in counseling. *Asia Pacific Journal of Counselling and Psychotherapy*, *2*(1), 25–40. https://doi.org/10.1080/21507686.2010.546865
Stueland, E. (2006). *Brått revet bort* Forlaget Oktober.
Suits, B. (2006). Games and their institutions in *The Grasshopper. Journal of the Philosophy of Sport*, *33*(1), 1–8. https://doi.org/10.1080/00948705.2006.9714686
Swedish Accident Investigation Authority. (2026). *Final report RL 2016:11e Accident in Oajevágge, Norrbotten County, Sweden on 8 January 2016 involving the aeroplane SE-DUX of the model CL-600-2B19, operated by West Atlantic Sweden AB*. File no. L-01/16 12/12/2016.
Symington, J., & Symington, N. (1996). *The clinical thinking of Wilfred Bion* (p. 198). Routledge.
The KLM-Pan Am Accident. (1977). https://aviation-safety.net/asndb/329178
UK Parliament. (n.d.). "Never in the field of human conflict was so much owed by so many to so few." (20 August 1940). Retrieved 19 August 2025 from https://www.parliament.uk/about/living-heritage/transformingsociety/private-lives/yourcountry/collections/churchillexhibition/churchill-the-orator/human-conflict/
Vaughan, E. J., & Elliott, M. W. (1978). *Fundamentals of risk and insurance* (2nd ed.). New York: Wiley.
Visser, M. (2003). Gregory Bateson on deutero-learning and double bind: A brief conceptual history. *Journal of the History of the Behavioral Sciences*, *39*(3), 269–278. https://doi.org/10.1002/jhbs.10112
Weick, K. E. (1995). *Sensemaking in organisations*. Sage Publications, Inc.
Weick, K. E. (2001). Leadership as the legitimization of doubt. In W. Bennis, G. M. Spreitzer, & T. G. Cummings (Eds.), *The future of leadership: Today's top leadership thinkers speak to tomorrow's leaders* (pp. 91–102). San Francisco, CA: Jossey-Bass/Wiley.
Wiener, E. L. (1988). Cockpit automation. In E. L. Wiener & D. C. Nagel (Eds.), *Human factors in aviation* (pp. 433–461). San Diego, CA: Academic Press.
Winnicott, D. W. (1971). *Playing and reality*. London: Tavistock Publications.
Wittgenstein, L. (1922). *Tractatus logico-philosophicus*. New York: Harcourt, Brace and Company Inc.
Wittgenstein, L. (1953). *Philosophical investigations*. London: Blackwell.
Wittgenstein, L. (1978). *Philosophical grammar* (R. Rhees, Ed.). University of California Press.
Wolfe, T. (1979). *The right stuff*. Farrar, Straus and Giroux.

Index

Pages in *italics* refer to figures.

For Product Safety Concerns and Information please contact our
EU representative GPSR@taylorandfrancis.com Taylor & Francis
Verlag GmbH, Kaufingerstraße 24, 80331 München, Germany